BIRDWATCHERS' YEAR

BIRDWATCHERS' YEAR

by Leo Batten, Jim Flegg, Jeremy Sorensen,
Mike J. Wareing, Donald Watson
and Malcolm Wright

With illustrations by IAN WILLIS
and DONALD WATSON

T. & A. D. POYSER LTD
Berkhamsted

CONTENTS

Publisher's note 7

The contributors 9

URBAN AREA *Leo Batten* 13

WOODLAND *Jim Flegg* 67

WETLAND *Jeremy Sorensen* 133

FARMLAND *Mike J. Wareing* 191

MOUNTAIN AND MOORLAND *Donald Watson* 243

ISLAND *Malcolm Wright* 301

PLATES

Urban area *facing pages 72 and 73*

Woodland *between pages 72 and 73*

Wetland *facing pages 176 and 177*

Farmland *facing pages 192 and 193*

Mountain and moorland *facing pages 288 and 289*

Island *facing pages 304 and 305*

PUBLISHER'S NOTE

Birdwatchers' Year is, in essence, six 'diaries' recording bird activity and behaviour in six differing habitats: the vicinity of Brent Reservoir in NW London; a Kentish woodland; the Ouse Washes in Cambridgeshire; farmland in Derbyshire; a mountain region of SW Scotland; and an island, the Calf of Man. This variety of habitats, climate and topography happily encompasses a wide and varied range of breeding, resident and passage birds.

The 'brief' for the sextet of authors was deliberately imprecise, requiring only an account, month by month, of a typical year in the chosen habitat— an account which need not necessarily confine itself to birdlife but which would include any relevant comment on plants, other animals, man and his intrusions, climate, and so on. As a result, each author's approach is different and individual, and the 'diaries' are diverse in style, content and viewpoint. Within *Birdwatchers' Year* comparisons are not odious, and it is interesting, in some instances, to compare authors' accounts of events during a particular month. However, the reader should remember that the accounts may not relate to the same year or, indeed, to any particular year. To have restricted the authors in such a way would have meant the omission of much that was interesting and noteworthy from observations and study spread over several years.

The common factor throughout the 'diaries' is the authors' respect and affectionate regard for the birds themselves, their intimate knowledge of the study area, and their evident understanding and knowledge of the birds and their environment. In the event, the resulting combination of fact and information, comment and bedside reading, and the evocation of seasons and places, will undoubtedly achieve the objective of giving pleasure and an extension of interest to a great many birdwatchers.

The drawings at the month openings are by Ian Willis throughout, except for those accompanying 'Mountain and Moorland' which are by its author, Donald Watson.

In a few instances the exact location of rarer breeding species has been disguised; and an index has been omitted—the number of references for many species would have been too numerous to have been meaningful.

THE CONTRIBUTORS

LEO BATTEN (Urban area)

Leo Batten has lived in NW London near the Brent Reservoir all his life and became interested in watching birds at the age of nine. The nearness of the Reservoir to his home stimulated his interest and now, although based at the British Trust for Ornithology at Tring, he retains a strong interest in suburban bird life and the problems of its conservation. He believes that careful planning of amenities in urban areas would enable a rich wildlife to be retained in pockets of open space, and lengthy discussions (in which he has been involved) with the local Councils have resulted in the creation of an urban environmental study area near the Reservoir. The study area will be used for teaching young children something about natural history in the hope that their understanding will lead to a greater respect for animal and other forms of life and the environment in which they exist. He believes that it is natural for young children to be interested in many aspects of natural history, but children living in high-density housing estates seldom have an opportunity to develop this interest, and an above-average proportion of them tend to have little respect for life or property, as the prevalence of vandalism shows.

Since joining the BTO he has worked on the Common Bird Census and is carrying out a detailed study of Blackbird population dynamics and roosting behaviour—much of the work being done around the Brent Reservoir. He lectures frequently and is involved in advisory work—most of it in the British Isles but also in Europe. He is the author of many scientific papers and popular articles on birds. He has studied birds in Europe, North Africa and Asia and is currently engaged in a book on roosting behaviour of birds.

JIM FLEGG (Woodland)

Dr Flegg first visited Northward Hill 25 years ago, as a schoolboy birdwatcher still in short trousers. Since that time, he has become more and more concerned with understanding the various natural factors at work within the wood, trying to interpret the observed biological changes for the general benefit of the woodland and its many birds. As an Honorary Warden of the Reserve for many years, as a research worker and as a birdwatcher, he spends as much time as possible in the wood—often more than fifty days each year. A zoologist by training, his ever-growing involvement with the *study* of birds (which, he argues, *increases* the enjoyment to be obtained from birdwatching) led him to abandon a career in research into agricultural pests and turn to birds. In 1968

9

he became Director of the British Trust for Ornithology, the society whose aim it is to coordinate like-minded birdwatchers in national or local co-operative studies of birds, part of whose efforts are at present directed towards understanding the impact of birds on man, and much more of whose efforts are devoted to assessing the effects of our rapidly changing environment on birds. The BTO is responsible for amongst many things bird ringing, census work and the atlas of the distribution of our breeding birds. Some of the practical applications of this sort of birdwatching become apparent in this account of the year in Northward Hill.

JEREMY SORENSEN (Wetland)

His general interest in nature dates from early childhood. Later, when at Leighton Park School, Reading, he joined the RSPB and the school bird club, and began to learn how to ring birds, soon becoming so interested in birds that there seemed little time for lessons. In 1952 he left school and started work with a Manchester firm of wall paper manufacturers but continued to spend most of his spare time birdwatching. He became a BTO member in 1954, and in that year was called up for National Service. Released by the Army in 1956, he returned to his previous employment. In 1959 he obtained his 'A' ringing permit and has since ringed more than 60,000 birds, mainly with Alan Burgess at the edge of the Pennine Moors in Derbyshire, Cheshire and Lancashire. In 1967 he was elected a member of the British Ornithologists Union and gave up his job to go on a six-week expedition to Spain and Portugal to study, under Chris Mead's leadership, trans-Saharan migrants (mainly warblers and flycatchers). In November 1967 he joined the staff of the RSPB and became assistant warden to John Wilson at Leighton Moss Reserve. In April 1968 he moved to the Ouse Washes as warden.

MIKE J. WAREING (Farmland)

Mike Wareing's family have been farmers for many generations, and no doubt this accounts for his natural interest in all forms of plant and animal life. He recalls going for walks with his uncle at a very early age (he remembers a Snipe's nest he was shown when only three years old), and it was from his uncle that he developed a keen interest in natural history, in birds in particular. This interest was further stimulated when, aged about nine, he found the body of a bird in one of the fields and learned, later, that it was a Black-throated Diver. At grammar school some of his enthusiasm waned due to studies and an increasingly serious involvement in sport and athletics. When obliged to give up athletics and turn to farming he took up farming politics and became active in the National Farmers Union. His developing political awareness, however, awakened him to the impact of man on the environment, and his interest in ecology and ornithology returned. Soon afterwards, he met David Blackmore who introduced him to ringing and encouraged him to start the Breck Ringing Group. His interest in ecology has also led him to take an active part as a member of the Derbyshire Naturalists Trust.

DONALD WATSON (Mountain and moorland)

Began to watch and draw birds in early childhood, in company with his brother, Eric. Enthusiasm for ornithology was fostered by growing up in Edinburgh as a member of the Midlothian Ornithological Club, with the delights of the Firth of Forth and its islands at hand. Graduation in Modern History at Oxford was followed by nearly six years as a wartime soldier and while in Burma he resolved to try to make a livelihood from painting landscape and birds after demobilisation. Since 1948 he has exhibited work widely, held several one-man shows and become a founder member of the Society of Wildlife Artists. More recently he has illustrated a number of books, including *Birds of Moor and Mountain* (which he also wrote), and the *Oxford Book of Birds*, and has contributed numerous drawings and occasional articles to magazines and journals. He is married with four children and lives among the hills of SW Scotland where for over 20 years he has made a close study of the birds. He is active in conservation and bird protection, through the Scottish Wildlife Trust and the RSPB, has been a Council member and is regional representative of the BTO, and was President of the Scottish Ornithologists' Club, 1969–72.

MALCOLM WRIGHT (Island)

Born at Kingswinford in south Staffordshire he began his bird-watching among the fields, woods and reservoirs of that area at the age of ten. On leaving school he went into accountancy and spent eight years in that profession, qualifying as a chartered secretary and certified accountant, but was never happy at being chained to an office desk while the sun was shining outside. In 1965, by which time he was working in London, he made his first visit to the bird observatory on Bardsey Island in North Wales and very quickly became an island addict. He took up with alacrity the offer of the assistant warden's post on Bardsey in 1966 and returned to that splendid island the following year. Since March 1968 he has been employed by the Manx Museum and National Trust as warden of the bird observatory on the Calf of Man. He enjoys the outdoor island life immensely and says that his favourite birds are Choughs, Peregrines and Stonechats.

URBAN AREA

by LEO BATTEN

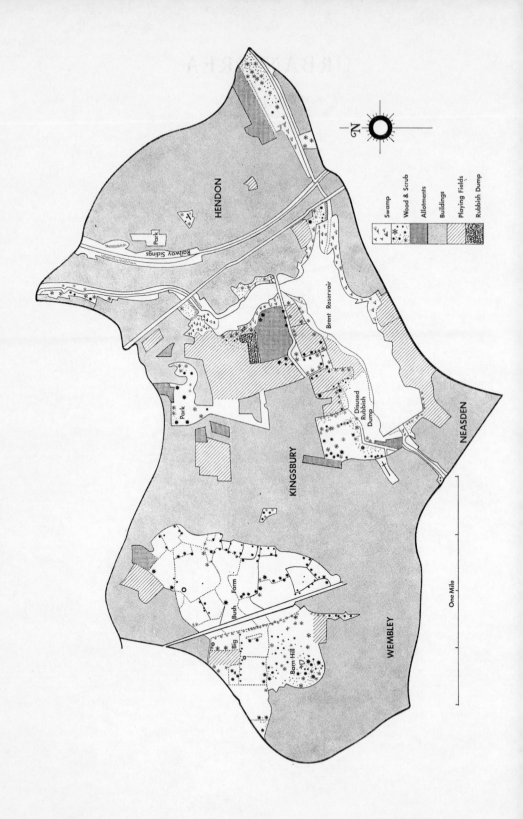

URBAN AREA

N

Swamp
Wood & Scrub
Allotments
Buildings
Playing Fields
Rubbish Dump

HENDON

Park

Railway Sidings

Park

Brent Reservoir

Disused
Rubbish
Dump

KINGSBURY

NEASDEN

Bush Farm

Big

Barn Hill

WEMBLEY

One Mile

Introduction

The study area is located approximately six to seven miles north-west of Marble Arch, and is bounded on the south side by the North Circular Road, and on the east by the Watford Way and Hendon Way. Kingsbury Road forms the northern limit joining with the Watford Way via Rushgrove Avenue and Colin Deep Lane. Finally the western perimeter follows Neasden Lane, Forty Lane and then along the Bakerloo electric railway to Kingsbury.

In all, the study area covers 817 hectares (2,018 acres), the highest point being Barn Hill at 282 feet from where the land slopes away to 100 feet on the banks of the Brent Reservoir and again rises to just over 200 feet in the north-eastern corner of the area. The ground in the centre also rises to 200 feet so that the whole area takes on the form of a shallow bowl raised at the centre.

History of the Area

The Brent Reservoir came into existence at the completion of a dam across a small part of the Brent valley in 1833. The Reservoir itself was not finished until 24 November 1835. Later a small extension was added, the whole construction of this canal feeding reservoir being finally completed on 15 December 1837.

This stretch of water and the surrounding districts of Kingsbury, Hendon and Barn Hill have received the attention of generations of naturalists since the moment the reservoir was formed. Frederick Bond, one of the founders of the *Zoologist* was living in Kingsbury at the time and is primarily remembered for the number of rare birds which he obtained at the Reservoir in the early years of its existence. Birds such as Squacco Heron, Night Heron, Little Bittern, Pomarine Skua and Temminck's Stint figure amongs this frequent records to the *Zoologist* at that time.

The variety of bird life which frequented the Reservoir in the early years is astonishing, but then it was, for the first ten years of its existence, in many different conditions. The first four years saw it as an expanding shallow stretch of water flooding wet meadow land. After a few years as a completed Reservoir a particularly stormy period in January 1841 resulted in all the fields between Neasden and Stonebridge being flooded from excess water from the Reservoir. For a whole week water poured over the floodgate with a mighty roar, which a chronicler of the time has recorded as rendering the chimes of Old Kingsbury Church almost inaudible. The rain continued and on the stormy night of 16 January 1841, the dam finally burst. The water swept on unimpeded and many people and much livestock were drowned. Once again the water level in the Reservoir was low and until the dam was repaired it would have been in a superb state to attract marshland species of birds.

15

It was Bond's influence which stimulated J. E. Harting to compile his book *Birds of Middlesex*, published in 1866. As much of the information in that book refers to north-west Middlesex, especially around the Brent or rather Kingsbury Reservoir as it was then known, it was the main source of information on the bird life of those times, when it was a rural beauty spot.

Bond died in 1888 and part of his obituary written by Harting contains a graphic description of the area in its heyday. 'Kingsbury reservoir was our happy hunting ground in those days (twenty or five and twenty years ago) it was a paradise for an ornithologist. There was no railway viaduct at one end of it then as now, the extension of the midland line to Bedford had not been commenced. When we visited London we had to drive our own horses, or go by one of the two coaches which were then on the road, one of them going to and from St. Albans, the other to Stanmore and Elstree. It was no uncommon thing as we crossed the two bridges over the reservoir and the Hyde Water to see Wild Ducks there, and gulls and terns flying about at the period of the migration in spring and autumn. About the end of April and beginning of May and again in August to about the middle of September the number and variety of wading birds which visited this fine sheet of water were most remarkable. Plovers and Sandpipers, Snipe and Jack Snipe, were all there in their proper season, and there were always a few Herons about which came either from Osterley Park, Black Park, Uxbridge or Wanstead Park in Essex. The water was very little disturbed then by human visitors and we have many a time walked around it, about two miles, and followed the Brent towards Hendon or in the other direction towards Brentford without meeting anyone but farm labourers, or perhaps one or two anglers. Here in the early morning might be heard the note of the Ringed Plover as it ran along the shingle at the head of the reservoir, or the musical cry of the high flying Redshank which we marked down, to be stalked and shot. On the muddy margins in the bed of the brook, especially at a bend devoid of trees, the Green Sandpiper might be found every spring and autumn, and more rarely the Wood Sandpiper and Temminck's Stint. We shot over four parishes, including a bare open tract lying between Kingsbury, Kenton and Edgware, known as Hungry Downs, where Golden Plover and Peewits came in winter and the Dotterel appeared in spring and autumn.'

The rest of the area at that time was mainly permanent pasture with scattered copses, overlooked by the wooded height of Barn Hill. Apart from the construction of the railway in 1868 which crossed the eastern extension of the Reservoir, little change in land use occurred until the 1890s when rows of houses were built close to the northern end of the Reservoir.

By 1913 urbanisation of the area had got under way and by then 10% of the land surface had been built on, mainly towards the north and east of the

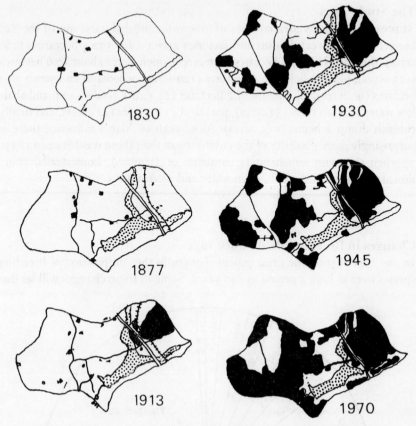

The study area at intervals, 1830–1970, showing built-up areas in black

Reservoir. Further development took place in the 1920s when the North Circular Road was constructed, and by 1930 30% of the area was under development. This figure had increased to 45% by 1945 and no less than 65% by 1970. The gradual increase in Urbanisation is illustrated above, showing the area at intervals since 1830.

By the 1960s a considerable amount of attention was centred around the Reservoir. In these later years apart from regular observations, organised census work, breeding biology, ringing studies and bird atlas work have been carried out, and for the first time quantitative ideas of the avifauna have been obtained. Major alterations to the area are now taking place in the form of new high density housing estates, replacing substantial houses and their large wooded gardens. Further changes are envisaged for the future and at this stage it is difficult to feel optimistic about the effects these will ultimately have on the bird life, but at least these effects are likely to be well documented.

The Study Area

At present the study area consists of nine main sub-habitats; out of the 817 hectares (2018 acres), houses and factories account for 530.4 hectares (1288 acres), the Barn Hill 97.1 hectares (240 acres), open water about 50.6 hectares (125 acres), playing fields 52.2 hectares (129 acres), woodland and scrub 36.4 hectares (90 acres), allotments 17.8 hectares (44 acres), reed swamp and shallow water 13.3 hectares (33 acres), parkland 17 hectares (42 acres), and finally rubbish dump 2 hectares (5 acres). As a result of Man's influence there is surprisingly more diversity of the environment than there was between 1833–77 when the main sub-habitats consisted of farmland, homesteads, stony downland, wood and scrub, open water and reed swamp.

Changes in Breeding Birds since 1833

In any area of land one must expect changes in the composition of breeding species over as long a period as 140 years. Some of these changes will be due

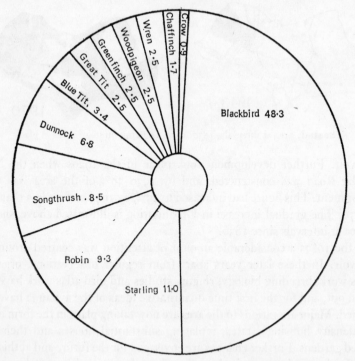

1. Suburban gardens—'pie' diagram showing percentage distribution of species in study area. (See page 34).

to extrinsic factors such as variations in climate or modifications to the area itself. Other intrinsic differences may be caused by long term and widespread population changes in the birds themselves. It is often difficult to be sure which of these factors is responsible for changes in status of certain species, particularly as there is a possibility that suburban habitats may be second-rate for some species, and occupied only in years of high population level. Long term population fluctuations which might occur nationally or internationally and go unnoticed elsewhere could be accentuated in this sort of area. Although no quantitative data exist, there is evidence based on presence or absence that the Green and Great Spotted Woodpeckers have fluctuated markedly in the last 140 years; they were described as common in the early decades of the nineteenth century, were rare in the latter half, and common again by the 1920s. They have been declining since the middle 1950s. The first oscillation occurred when the area was changing little, in fact the Green Woodpecker was absent as a breeding bird for some 30 years and the Great Spotted Woodpecker for nearly 40 years.

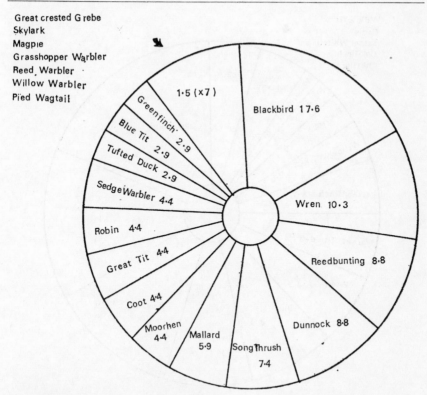

Great crested Grebe
Skylark
Magpie
Grasshopper Warbler
Reed Warbler
Willow Warbler
Pied Wagtail

1·5 (×7)

Greenfinch 2·9
Blue Tit 2·9
Tufted Duck 2·9
Sedge Warbler 4·4
Robin 4·4
Great Tit 4·4
Coot 4·4
Moorhen 4·4
Mallard 5·9
Song Thrush 7·4
Dunnock 8·8
Reedbunting 8·8
Wren 10·3
Blackbird 17·6

2. Willow swamp in the study area—'pie' diagram of species present. (See page 38).

During the period under consideration a number of other species have fluctuated in such a way as to show a return to a former status after a period of scarcity or abundance. The most noticeable of these are Coot, Stock Dove, Nuthatch, Goldfinch, Bullfinch, Redpoll, Jay and Magpie. As some of these population changes took place when the area was not being extensively urbanised it seems likely that other factors were largely responsible for the early fluctuations. Other species such as Grey Wagtail, Reed Warbler, Night-jar, Pheasant, Tree Pipit and Cirl Bunting may have been affected similarly. Red-legged Partridge and Little Owl have also had a period of regular breeding in the area but these were both species introduced into the country in the nineteenth century.

A number of species declined and disappeared as breeding species fairly rapidly by the end of the first decade of the present century. Although only about 10 per cent of the land was urbanised by then, the surrounding districts nearer London were being developed and with this came the inevitable increase in disturbance from the human population. Some species lost to the

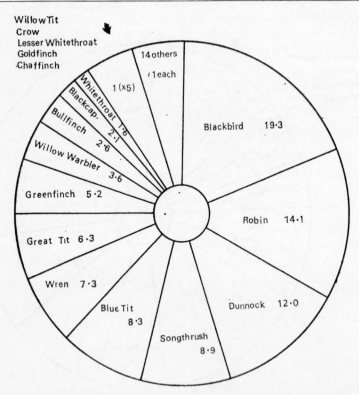

3. Woodland and scrub—'pie' diagram of species present. (See page 41).

area at that time were high forest species which appear not to be able to tolerate human disturbance. Hawfinch, Redstart and Wood Warbler are in this category. The Kingfisher declined as a result of the increased pollution of the rivers and tributaries in the area. The Lapwing and Corncrake were also lost and it is likely the former went because of the continual plunder of its eggs. The decline of the Corncrake is a mystery, and the loss of Goldcrest and Nightingale seems premature.

On the credit side the Great Crested Grebe appeared, presumably helped by the Bird Protection Act of 1880 which banned its slaughter to adorn Victorian ladies' hats. This fashion did not die out until about 1908. The Coot also reappeared but at the time of writing there are signs of a decreasing breeding population and an extremely poor productivity rate at the reservoir, so this species may once again disappear as a breeding bird in the not too distant future. The provision of nesting rafts this year enabled more Coot to be reared and may prevent the loss of this species as a breeding bird at the Reservoir.

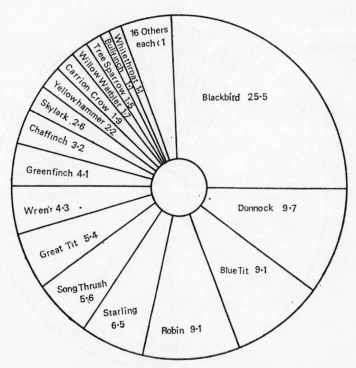

4. Study area farmland—'pie' diagram of species present. (See page 44).

Of the breeding species which have been lost to the area at least for the time being nearly 60% have declined nationally and this probably has contributed to their disappearance. This leaves Turtle Dove, all three woodpeckers, Kingfisher, Swallow, Meadow Pipit, Wood Warbler, Nightingale, Long-tailed Tit, Nuthatch, Treecreeper, Goldcrest, Hawfinch and Jackdaw which are presumably victims of the changes inflicted by urbanisation.

It is clear that the number of species present in the breeding season has decreased as the area has become increasingly built up. Although the total number of species regularly breeding has dropped from 72 to 43, only 37 of the species which bred before 1860 still do so. This means that six regular breeding species have been gained—Little Grebe, Mute Swan, Tawny Owl, Magpie, Swift and Tree Sparrow. In addition to this a further three species are now recorded as occasional breeders but were unknown as such before 1860. These are Great Crested Grebe, Tufted Duck and Grey Wagtail.

Detailed information on species present in each breeding season since 1957 is available. During this time the area has been 60–65% urban. There is

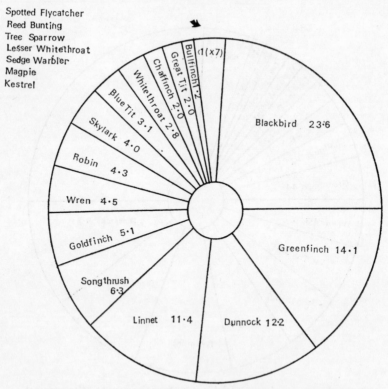

5. Study area allotments—'pie' diagram of species present. (See page 45).

considerable variation in the actual species involved each year. For example, out of 52 which bred in at least one of the years only 25 (48%) bred every year; and out of the 66 species which were observed in at least one breeding season only 37 were seen in all the years concerned. It would seem therefore that the breeding avifauna of this urbanised area contains a very large proportion of irregular breeders.

JANUARY

My first visit to the area this year was made on the 2nd January. When I arrived an hour or so after dawn the sun was already bright in the sky, there was no wind and it was uncannily warm for the time of the year. Out on the reservoir there were over a hundred diving duck, 60 Pochard, half as many Tufted Duck and 13 Smew. As the sun shone on their plumage I could not help comparing the appearance of the reservoir and its surroundings with the scene nine years earlier, during the great frost of 1962–63.

By 2nd January 1963, the reservoir looked like a scene from the arctic with snow up to two feet in depth and ice covering the reservoir three inches thick, except for a small pool in which were crammed the 30 or so remaining ducks, together with another 30 Coots. A few days previously there had been some spectacular overhead cold weather movements, starting in earnest just after Christmas. On 28th December nearly 1,000 Skylarks flew west in four hours of very chilly observations; hundreds of Woodpigeons could be seen passing west and there had been a large influx of finches and other passerines in the area, including 12 Woodlarks and a Snow Bunting. The birds just kept pouring over without ceasing, flocks every now and again dropping in to feed briefly and then fly on. Large flocks of Goosander and White-fronted Geese both rarely seen in the area were amongst the fleeing birds, clearly we were in for a lasting spell of bad weather, but even then we did not realise just how bad.

The overhead passage reached a maximum on the 5th January when 1600 Fieldfares and 700 Redwings flew west. Woodpigeons were also recorded in hundreds but these were passing over in all directions. Another flock of Goosander, only 18 this time, flew over and finches and buntings started appearing around the reservoir feeding on the husks of any seed bearing plant still not completely covered by the snow. These were to be succeeded by huge nomadic flocks of Skylarks, hundreds strong, desperately searching the ice for windblown seeds which must have been more visible on the frozen reservoir— ice that was by the middle of the month six inches thick. A Sparrowhawk paid a fleeting visit scattering the Skylarks on the ice but making no attempt to catch any, just carrying on down the reservoir presumably on its way south hoping for warmer weather. By the end of the month most of the birds had evacuated the area, those that remained had taken to the gardens and the rubbish dumps—Fieldfares, Redwings, Starlings, together with Chaffinches, Tree Sparrows, Blue and Great Tits, Robins and Dunnocks. Dead and dying Coots and Moorhens lay everywhere, they had not had the good sense to get out whilst they could.

So it was to continue for another six weeks when the only common birds were gulls riding out the frequent blizzards and up to 24 Fahrenheit degrees of frost as the area continued to disappear under more snow so that eventually there was no way of distinguishing the reservoir from the surrounding land.

Yes, these two winters nine years apart are about as different as they could be. This year there has been not the slightest suggestion of overhead movement and although gardens with a regular food station are receiving a steady stream of birds, most birds have chosen to remain independent of man and seek their food elsewhere; the thrushes on the playing fields, the finches and buntings on the farm or rubbish dumps, whilst the scrub holds the Robins, Dunnocks and tits.

One has to get to the reservoir early on a Sunday morning to be there before the emergence of the sailing dinghies, because most ducks leave on their appearance and today was no exception; the main flock of diving duck were off as soon as the first sail appeared around the headland and the remaining birds went when the boat was about 400 yards away. Fortunately the Smew are not so boat conscious and flew up to the eastern end of the reservoir. This part is rapidly becoming silted up with numerous small muddy islands appearing above the water, these become rapidly vegetated with willows and reed grass and provide a good refuge for waterfowl. In fact, the Smew spend most of their time here feeding in the shallow water. It is not certain what it is that they find so attractive but some of the spots where they feed are only a few feet deep and the muddy bottom contains millions of small molluscs particularly *Limnaea pereger* which in places cover much of the mud surface. Perhaps these are the big attraction for the Smew at this disturbed grubby urban canal reservoir. The Brent is no doubt one of the top five areas for this species in the country. For a while in the mid 1950s it was the most important area supporting up to 144 Smew in the winter but now the wintering flock seldom tops 35 birds.

The biggest threat to the Smew is the loss of this last refuge at the reservoir. Boats have always been sailed on the main stretch of the water, in fact the first boat was on the reservoir while it was under construction in the late 1830s. In the early 1960s sailing was also allowed on the smaller arm, and now with increasing demand for water sports there is talk of a boating pool at the marshy eastern end, that is after the reservoir has been dredged by the Greater London Council and the fertile silt deposits at the eastern end removed for flood alleviation purposes. The area is a Site of Special Scientific Interest and although that name tag affords very little protection to an area, it is hoped by the local conservationists that when the dredging takes place the marsh will not be lost but improved and discussions have taken place with the appropriate councils to this effect. The situation is hopeful. The ideal would be the creation of a wetland reserve with hides and a nature trail to interest some of the people who walk around the reservoir.

The weather is still mild now in the middle of the month and some short bursts of song have been heard from Mistle Thrushes and a few Blackbirds. There are very few Redwings and no Fieldfares at the thrush roost this year, even the Blackbirds are fewer in number. We have been ringing and weighing Blackbirds at this roost now for nearly ten years and find that the winter weights of birds caught in the area during cold spells with snow cover are usually higher than the weights of birds caught in more rural surroundings. This suggests that winter (except the 1962–63 winter) is a less difficult time for suburban birds than for rural ones, presumably because of the abundance of food stuffs put out for them. As an example the average weight of sixty-four Blackbirds caught in the first fourteen days of January 1969 was 125.0 grams, indicating about 20 grams of fat and enough for nearly three days survival without more food. At Northward Hill, High Halstow in Kent during the same period the average weight was only 92.5 grams indicating about 7 grams of fat, sufficient for less than one day's requirements. Almost all the Blackbirds at the Brent roost are from the surrounding gardens and few travel further than three quarters of a mile to the roost which is sited in a dense hawthorn thicket.

When the weather is mild for the whole of January as it is this year, very little spectacular tends to happen. The Smew population remains low, in fact there are only 10–14 of these birds present now; there seems to be hardly any in the country this year. At the end of the month a flock of 30 Reed Buntings were found feeding on the Reed grass. We have never had so many together. The spring passage does not take place until late February and early March; if these birds are passing through they are a full month earlier than usual. The

Reed Buntings could not be found on 29th but we had a new visitor to keep us occupied; a Bearded Tit had taken up residence in our tiny Phragmites reed bed. We rarely get these superb birds here although last October I had the good fortune to see 17 together in the reeds by the reservoir, and was able to catch and ring six birds. Two were later caught by Jeremy Sorensen at Manea in Cambridgeshire along with some Minsmere ringed birds. It is only in the last 12 years that irruptions of Bearded Tits from East Anglia and Holland have got as far as the London area in the autumn; before then they were quite unknown. To see 17 in an area only seven miles from St Paul's Cathedral was really fortunate.

FEBRUARY

The weather is still very mild for the time of the year so I was surprised to find a huge flock of diving duck at the reservoir on the 5th, no one had mentioned unusual numbers yesterday; these must have flown in overnight. We spent some while closely examining each one but the entire flock of 295 consisted of 140 Pochard and the rest Tufted Duck. They were stretched in a thin line right across the reservoir, most were loafing, and a few were feeding, that is until the dinghies appeared. Then, the entire flock took off and split into three smaller groups as they wheeled and circled over the reservoir like flocks of waders with their light bellies visible one second and their dark backs the next. It was quite a spectacle and we watched them until they were mere specks in the sky, heading in a northerly direction probably going to Hilfield Park reservoir seven miles to the north. The reservoir suddenly looked deserted except for nine Smew and some 40 Mallard drifting up to the secluded eastern end.

The next week passed with little excitement, the Bearded Tit had moved to a large clump of reed grass at the northern end of the reservoir where it afforded good views to a number of bird watchers. It seems unlikely that we will get any cold weather now to speak of yet despite the mild conditions I have not seen a single butterfly this year. In 1959 when we also had a warm February there were dozens of Small Tortoiseshells all over the area by now. Still there is a very nice show of Coltsfoot, particularly in the proposed nature reserve area near the western end of the reservoir.

An unusual visitor on 19th February was a Lesser Spotted Woodpecker.

There is an increase in this species in southern England, and it has been suggested that the rampant Dutch Elm disease has opened up a good food supply and provided a temporary abundance of nesting sites. We appear to be lucky so far, most of the large elms at the Brent showed no visible signs of the disease last year, although one specimen over 100 years old died from it as did several 30 year old trees. The trouble is one cannot tell the current state until the signs of withering leaves become visible in July. The landscape is going to look very different here in a year or two if the disease catches hold. The elm is the dominant tree species and there must be dozens around the reservoir which are over 150 years old. One can only hope they will be strong enough to withstand this virulent fungus.

In this area of London there is an ever increasing demand for developing what is left of the open ground, so much so that scarcely a few months go by without a new application for building being received. Some of them have been successfully appealed against, others have gone ahead with some sort of compromise situation for the wild life interests of the area. As there was little of interest happening we decided to take the opportunity to see if the contractors who are building a road to the factories have kept to our compromise plan to create a minimum of disturbance to the eastern marsh.

This particular encroachment started last year with a planning application for a small access road to the British Oxygen Company. Their front access route on the North Circular Road will be closed with the extension of the M1 near the Edgware Road. There seemed little hope of stopping the access road being constructed, even though it cut off a corner of the eastern marsh. No sooner had this application been agreed than a further one was received from the contractors to use some adjoining ground then covered with developing hawthorn scrub for storage of top soil and siting of temporary office space. Their proposal was to clear the vegetation right to the reservoir banks. After some discussion we managed to get an agreement they would leave a buffer zone of vegetation of some 40 feet from the reservoir banks, behind which they would pile the top soil and site their temporary offices nearest the Edgware Road. So far the buffer zone has been maintained and virtually no disturbance has resulted to the marsh. However, some two acres of scrub have been destroyed and one wonders what is next to go.

The pleasant weather at the end of the month enabled us to get down to some of our plans for improving the marsh. Most of the reservoir is surrounded by a narrow reed bed. Unfortunately the predominant grass is *Phalaris arundincea* which is rather inferior to *Phragmites communis* as a bird habitat. The stems of *Phalaris* are normally too weak to provide an anchorage for nests. We do, however, have a slowly expanding bed of *Phragmites* in the eastern marsh. Our object was to extend this. We took the rhizomes from the original bed and transported them across the River Brent which bisects the marsh, and planted them on an area of recently uncovered mud. We dug holes about a foot deep and five foot apart and popped a few rhizomes in each. There seems some controversy about when to plant *Phragmites*: autumn planting may be best as it gives the plants time to settle down, on the other hand they can easily be washed out of the mud by the winter floods. I doubt if we will see any growth of the transported reeds above the mud this year as they have been planted so late.

Our other task was the construction of nesting rafts. With the increasing concrete coverage of the surrounding land surface, less and less rainwater can drain away through the soil. The result is after heavy rain the small rivers and tributaries feeding the reservoir become swollen torrents from the water washing off the roads; some times the River Brent can rise four feet in as many hours. In these circumstances the shallow reservoir can also rise over a foot, flooding the surrounding vegetation and washing away a large number of nests of Coot and Moorhen. It has been mainly due to these spring floodings that no young coots have been reared at the reservoir for several years. Floating rafts were clearly needed for nesting sites.

Being an urban reservoir fed by urban rivers running part of their way through industrial areas we had a lavish supply of building material in the form of gallon petrol cans, old doors and rafters, new supplies of which regularly floated down the rivers into the reservoir.

We had no boat, in fact only a punt would have been practicable in the willow swamp. The water is often only inches deep over the rapidly forming silt banks which are extending further out into the reservoir every year. First a raft had to be built that would be strong enough to hold someone's weight without sinking more than a few inches into the water. It also had to be stable as the silt banks are up to six feet deep and very soft, so that falling off a raft could be quite dangerous.

After completion the nesting rafts were towed out one by one to the selected positions. The anchoring was tricky as it was necessary to transfer from the main raft to the nesting raft, a procedure watched with a great deal of amusement by those left on the reservoir banks. The final touches were then completed, tufts of reeds were attached to the bottom of the raft and several large sprigs of willow pushed through holes in the raft to the water beneath to provide anchorage for nests. (Willow has great powers of regenerating and the leaves opened on these sprigs at the same time as those on the growing willow trees, and remained green the whole season.)

All in all five rafts were constructed, the problem remained of what to do with the original raft. In the end it was decided to use it as a nesting raft even though whoever took it out would have to come back to the bank under their own steam. Fortunately one of the party in thigh waders had earlier lost his balance in the shallows and was well and truly saturated so he felt he had nothing to lose as long as he could anchor it over relatively firm substrate. This was accomplished without incident and we all left hoping our hard work would be taken advantage of by the waterfowl.

MARCH

One of the more unusual spectacles for Britain, which can sometimes be observed at the Brent Reservoir during the winter months, is the group display of the Smew. Prior to the exodus of these birds to their breeding grounds in Northern Scandinavia and Russia, pairing takes place. The ceremony begins suddenly when a small group of three to four males suddenly moves towards one female. The males jerk back their heads, at the same time raising their crests; this is followed almost immediately by a jerky sideways head movement and more crest raising. Occasionally a male will actually rear up on the water for a brief period. If the female receiving the attention does not respond by bobbing her head she is quickly passed by until the males find one who will. Strangely enough mating does not seem to take place after displays started by the male. Copulation occurs when the female solicits the male. This she does by swinging her bill up and down in an arc, the movement ending with the bill pointing towards the water. Sometimes she very spectacularly surges forward at a male with her head and neck stretched out in front while splashing up water with her feet. If the male does not take flight the female will present herself by sinking low in the water and mating may take place.

With the mild weather we lost our Smew early this year, all of them being gone by mid-March to begin their long journey back to their far northern breeding grounds. In cold winters they grace us with their presence up to the beginning of April. The Smew are amusing to watch but they cannot compare with the masters of the mating game, the Great Crested Grebes. Sadly we only have one pair of these handsome birds breeding with us now, and these usually attempt to nest in the Eastern Marsh. On most days about this time of the year, if one is patient enough, they can be seen indulging in their impressive head shaking display; both partners face each other, stretch up their necks, raise their crests and waggle their heads, at the same time gently swaying their necks; every now and again each bird turns its head to gently preen one of its scapular feathers and then returns to head shaking. A few years ago I was fortunate enough to witness the Grebe's fantastic Penguin dance at the reservoir, a rarely performed ceremony which once seen is never forgotten. Both birds dive to collect a piece of weed, they surface and swim towards each other dangling the weed in their bills, this action is not infrequent but usually it never gets further than head shaking. On this occasion they performed properly and when they were almost touching both birds reared right out of the reservoir, vigorously treading water to keep bolt upright. This lasted no more than a few seconds before they sank back to carry on with the more mundane head shaking. Strangely these ceremonies do not result immediately in mating, this takes place later on the nest.

The second half of March is rather like an interval in a play, one in which many of the actors change as well as the scenes. The winter visitors leave and early morning movements of Fieldfares, Redwings and Chaiffnches are a regular feature. The exodus is carried out quietly, without a fuss leaving the stage free for our summer visitors to take over. Great subtle transformations take place, the willow catkins burst forth, a billion buds prepare to open, spring flowers begin to carpet the untrampled ground and one becomes increasingly aware of the bird song which now fills the air. All this is happening within a few miles of the heart of London.

Encouraged by the warm sunny spell in mid-March this year we made a tour of inspection of our nestboxes in the wood. Fresh material brought in by Blue Tits was already in two of them—the beginnings of nesting activity. Everything seems to be forging ahead this year with the beautiful weather, lots of small Tortoiseshell butterflies have been tempted out and are flitting everywhere. I even saw two specimens of the strange ragged-looking Comma butterfly, a scarce insect in these parts. The volume of bird song after the mild weather is incredible, especially in the wood. The dawn choruses this last week have been some of the most remarkable I have ever heard. In this area the chorus is dominated by the Blackbirds, whose combined efforts produce a sheet of sound so rich it is not possible to distinguish the individual birds at all. One can look across the woodland scrub to the mass of chimney pots and television aerials and see dozens of small black blobs each proclaiming its territory. Above this chorus some louder but more beautiful songs are evident; these come from the scarcer Song Thrushes, and from the more immediate vicinity of the scrub the gentler songs of the Robin and Dunnock are added to this dawn activity, whilst the bell-like notes of the Great and Coal Tits contribute to the grand variety of sound. The number of birds singing in this period of free-for-all is so great that it is nearly impossible to census them.

For several years now we have been investigating the relative density and community structure of breeding birds in the various habitat types which make up this part of suburban London. The mapping technique employed in the BTO's Common Birds Census is used and this involves making ten or more visits to each study area during the breeding season. The census is carried out by plotting the positions of every bird seen or heard onto copies of 25 inch Ordnance survey maps of the area, a new copy being taken on each visit. The position of the birds is indicated by a letter code, e.g. B for Blackbird, R for Robin, Wr for Wren. If the bird is singing, the code letter is encircled. Song registrations are very important in this work as they indicate the position and approximate size of each male's territory. Other clues to the position of the bird's territory are also used such as alarm calls, young calling from the nest, birds collecting food or feeding newly fledged young. If several birds are heard singing in competition at the same time the registrations are joined together by a dotted line.

Every visit is given a letter, A, B, C, etc., in date order and the registrations of each species are transferred to another map of the area reserved for that species. For example, Blackbirds on visit A are taken from the visit map and plotted on the species map in exactly the same position. At the same time the symbol, B, for Blackbird is changed to A, to show on which visit the bird was seen. Similarly for all other visits. At the end of the exercise, clusters of registrations appear on the species sheet, each cluster representing the range of one bird's territory. By adding up the total number of clusters for each species we can construct a diagram showing the composition of the bird life in the area.

Census work has so far been carried out on the 240 acre farm at Barn Hill, a 24 acre area of scrub and woodland, 49 acres of gardens, 29 acres of allotments, 29 acres of willow swamp and 5 acres of *Phalaris* marsh.

The suburban garden census plot fell midway between the extremes of high density housing estates with small gardens which are found in the eastern end of the area, and the spacious large detached and semi-detached houses with ample, well wooded gardens in the southwestern section. Only thirteen species were detected as breeding on this census plot, not unexpectedly the House Sparrow being dominant. The Blackbird was the second commonest species followed by Starling, Robin, Song Thrush and Dunnock. The Blackbird, Starling, Song Thrush, Greenfinch and Wren are all typically woodland birds yet they exist at a higher breeding density here than they do in their ancestral woodlands. (See diagram 1, page 18).

APRIL

The mild spell of March has now ended and we have our usual cold dull weather which has been typical of English springs for some years now. Despite this, the dawn chorus from residential gardens and other open spaces, still dominated by Blackbirds is as impressive as ever. Surprisingly, no summer migrants turned up in March apart from one Wheatear on the 25th. Chiffchaffs and Willow Warblers were singing however by the 10th April, and Blackcap and Lesser Whitethroat were present by 15th. Some years ago I came across a note on the arrival of summer birds at Kingsbury, which is in our study area, in the *Zoologist* for 1844. It was interesting to compare the species involved then and the dates of arrival of those which are still regular passage migrants or breeding birds in the area. The table overleaf lists these birds.

Although the dates vary quite a lot, there is a tendency for those species which turn up at all to arrive slightly later now. Wryneck, Nightingale, Corncrake, Wood Warbler and Red-backed Shrike were all common summer residents in those days, whereas today they are quite unknown, even as spring migrants. Furthermore, only ten out of the original 22 species still breed regularly. Our only gain is Reed Warbler, but even this bred in most years about that time.

Our census work shows that the number of pairs of these summer visitors fluctuates sometimes quite markedly from year to year. None of these fluctuations, however, are ever as dramatic as the great crash in the Whitethroat

EARLY ARRIVAL DATES FOR 1844 AND 1957–66 (AVERAGE)

Species		1844	B = bred	Average 1957–66	B = breeds regularly b = breeds occasionally N = no longer occurs
Wheatear	March	25	B	March 24	
Chiffchaff		29	B	April 4	b
Wryneck	April	3	B		N
Willow Warbler		6	B	10	B
Blackcap		8	B	23	B
Garden Warbler		12	B	17	B
Redstart		12	B	16	
Tree Pipit		12	B	26	
Nightingale		12	B		N
Sedge Warbler		17	B	27	B
Swallow		17	B	17	
Whitethroat		20	B	17	B
Sandmartin		20		11	
Yellow Wagtail		20	B	15	B
Whinchat		20	B	20	
Lesser Whitethroat	April	22	B	May 4	B
Corncrake		22	B		N
Wood Warbler		22	B		N
Grasshopper Warbler		23	B	April 26	b
Cuckoo		23	B	May 5	
House Martin		27	B	April 25	B
Spotted Flycatcher		29	B	May 23	B
Common Sandpiper		29		April 19	
Red-backed Shrike	May	2	B		N
Swift		6		April 26	B
Reed Warbler		?		May 4	B

population in 1969 which reached international proportions. Prior to that year Whitethroats were just about the commonest warbler in the area. On our census plots in 1968, for example, we had a total of 24 pairs. In 1969 the number was reduced to three and there has been no improvement since then. We are still not certain of the cause of this widespread population crash. We do know that good numbers of Whitethroats left England in 1968, but they did not return. It has been suggested that mass mortality occurred because the rains, which usually come in February in the semi arid belt south of the Sahara, came early in December 1968. When the migrants arrived to fatten up for the long journey across the Sahara everything had dried up and there was little food. Those birds that did make it across the desert met with awful weather conditions in North Africa and the Mediterranean, and further heavy mortality took place. Not too surprisingly, the breeding population decreased by nearly 80% over the whole country and nearly 90% in this area.

April is usually a pleasant and varied month with a good deal of visible migration in evidence. The last Redwings and Fieldfares linger briefly in the area on their journey to their breeding grounds further north, whilst summer migrants appear in increasing numbers as the month progresses; Yellow Wagtails on the banks of the reservoir, Swallows, House Martins and Sandmartins dipping for flies over the water, whilst Sedge Warblers appear in the reedbeds and Willow Warblers and Chiffchaffs sing from the willows and the few remaining acres of woodland in the area. Sometimes large falls of Wheatears occur, as many as 17 together have been seen on one if the many playing fields; their favourite resting place. Terns are not infrequent at the reservoir and waders such as Redshank and Common Sandpiper are seen most years.

Shelduck, Common Scoter and Garganey may appear on the water and stay for a few hours, together with small numbers of Teal, Wigeon and Shoveler. The winter Tufted Duck and Pochard flock dwindles as the month progresses until only the summering, non-breeding individuals are left. At the same time small flocks of Great Crested Grebes appear, presumably on their way to their breeding quarters.

Until the last few years one of the most noticeable features of the spring was the large numbers (up to 500) of Lesser Black-backed Gulls which appeared at the reservoir. The numbers suddenly dropped off, together with other gull species, and this coincided with the closing down of a rubbish dump adjacent to the area, which provided good feeding for a large gull population. Small numbers of Lesser Black-backed Gulls still turn up and they are still much commoner than they were in Harting's time; he knew of only a handful of records up to 1866 when he wrote his book on the birds of Middlesex.

Much of our census work is carried out in April and May; the reed swamp areas tend not to be worth censusing until mid-April and need attention until mid-July, for some of the water birds are very late breeders. As I have mentioned, most of the reservoir is surrounded by a narrow reed bed. At the northern and eastern ends the marshy ground is rather more extensive. The eastern site, which is censused, owes it origin to the dry summer of 1959 when vegetation secured a stronghold on the exposed silt banks which have been formed by the River Brent over the last thirty to forty years. The silt being extremely fertile supports a lush vegetation dominated by reed grass and willows. Some of the latter are now over thirty feet high. There are a number of sheltered pools and an expanding *Phragmites* reed bed of about one acre, as well as extensive patches of the Greater Reed Mace. A slight decrease in the toxicity of the pollution from the adjacent factories over the last decade has also enabled vegetation to colonise the banks and the reed bed to expand. Part of this reed bed grows in a thin layer of humus and debris over a band of crystalline factory waste. The census plot also consists of about twelve acres of shallow water with many tiny islands of willows, the whole area of 29 acres being bisected by the River Brent.

Although occupying a very insignificant proportion of the total land area, these habitats are very important because a number of species are not found elsewhere in the study area. All the Moorhens, Coots, Little Grebes, Great Crested Grebes, Tufted Ducks and Reed Warblers breed or hold territory here, as well as most of the Sedge Warblers and Reed Buntings. An unexpected member of the marshland avifauna in 1970 was a Grasshopper Warbler which held territory in a bed of reed grass with scattered hawthorns and willows. In 1970 twenty-one species were found to be holding territory in the eastern marsh totalling 68 pairs, the commonest being Blackbird followed by the Wren and Reed Bunting. Diagram 2, page 19, shows the percentage of the total number of pairs each species accounts for in the eastern marsh.

MAY

From the air, the vicinity of the Brent Reservoir with its varied habitat in a desert of buildings is no doubt a considerable attraction to migrants passing over a heavily built-up area. From mid-March until early November many thousands use it as a convenient resting and feeding stage in their journeys. The reservoir itself is possibly the main attraction, and thereafter the various species soon disperse to their preferential habitats: the warblers to the thick cover, Yellow Wagtails, Whinchats and Wheatears to the grassland, and hirundines, waders, terns and ducks to the water.

The first week in May is one of the most exciting times of the spring migration as the largest falls of migrants occur at this time, generally after cloudy nights with rain and a south-east wind. If the water level is low then the shingle and muddy banks attract small numbers of waders. Green and Common Sandpipers, Little Ringed Plovers, Dunlin and Turnstone were all to be seen at the reservoir in the first week of May 1960 when the water level was lower than in any other recent spring. Black Terns usually turn up in the first two weeks of the month when there is an easterly wind, numbers are small with the maximum at any one time being 26 on 11th May 1960.

In more recent years the opposite condition, flooding, has been more usual and has resulted in many Coot and Moorhen nests being destroyed. It was pleasing to see that all the rafts constructed for waterfowl to nest on have been occupied. A pair of Scaup took up residence at the reservoir in April and May, and right up to 11th May defended one of the rafts from other birds. It would have been too much to hope for them to stay and breed here.

Bird song is also at its maximum now with virtually every breeding species

39

at the height of its nesting activities and a bright fresh early morning census visit to the wood can be very rewarding. With up to thirty species calling or singing, it is difficult to believe one is in a highly urbanised district only seven miles from Marble Arch.

The area of woodland censused was at one time destined to be a cemetery. In 1965 Willesden and Wembley were amalgamated to form the London Borough of Brent. There was a large cemetery with plenty of room in Wembley and therefore this area was no longer needed. Before the amalgamation was foreseen all the necessary preparations had been completed, including the construction of a chapel, a shelter and work accommodation for the cemetery staff.

The rest of the area consists of a wide belt of mature elm and oak bordering dense hawthorn, blackthorn and oak scrub up to 25 feet high. There are also patches of gorse, probably relict. A small wooded stream borders on one side but is, unfortunately, almost devoid of animal life owing to pollution caused by effluents from as yet unknown sources, and from oil washed in from the roads bordering its banks further north. This stream, and one other side of the wood, are bordered by houses, the remaining sides of the wood by roads and open fields. Adjacent to the wood is a long-disused playing field which has a wide belt of mature trees and scrub along one side. The field itself is rapidly becoming colonised by regenerating elm.

The site is very rich in wildlife, generally, including Moles, Foxes, Grey Squirrels, Weasels, Short-tailed Voles, Pigmy Shrews and Hedgehogs. There is also a good variety of Hymenoptera and Lepidoptera. It was here that the first Essex Skippers for Middlesex were recorded and these have since been shown to be as common as the Small Skipper.

The variety and density of breeding birds in the wood is remarkable. Excluding Woodpigeons, Starlings and House Sparrows which are very difficult to census by the mapping technique, there are on average some 210 pairs of birds holding territory in the twenty-four acre census plot. This is one of the highest densities recorded in any British woodland type, whether native trees or exotic. Thirty-two different species have bred or held territory in the wood in the last five years. The dominant species, not unexpectedly, is the Blackbird with 37 pairs, followed by Robin, Dunnock and Song Thrush with 27, 23 and 17 pairs respectively. The commonest warbler is the Willow Warbler but Blackcaps, Garden Warblers, Lesser Whitethroats and Chiffchaffs also breed. Whitethroats did so, too, until the population crash. Predators include Kestrel and Tawny Owl; unfortunately the Little Owl has been lost to the area within the last decade along with the Green and Great Spotted Woodpeckers. A pair of Whinchat bred in the wood in 1960, and in 1964 the Grasshopper Warbler did so; these species being amongst the nearest to the centre of London. Other species which have held territory recently include Pheasant, Spotted Flycatcher, Redpoll, Willow Tit, Yellow Hammer and Reed Bunting. A Cuckoo usually turns up for a few days each May but a female has not been heard for some years now. See diagram 3, page 20, for community structure.

Why the bird density should be so high is obscure, however it is typical of suburban woodland to have high densities. It is known that many birds use the surrounding gardens for feeding, and the diversity of the habitat provides a large variety of food. There are for example several acres of lawns nearby which are ideal for thrushes.

The largest thrush roost in the vicinity of the reservoir is located in dense hawthorns in this wood. This roost is in occupation throughout the year although, in summer, ringing has shown that most of the occupants are male birds. Some of these are territory holding males as they have been seen to stop singing at dusk and fly off to roost communally. The odd females which are at roost are presumably those which have lost their eggs or young to predators or are between clutches.

This outstanding area has been under threat of destruction in recent years by housing development. At present the threat has been removed, and instead it is proposed that part of the area be used for an educational field centre; just how much, is still under debate. A meeting is to be held later in the year when the final plans will be drawn up.

Each year during the last weekend in May a regatta is held at the reservoir with sailing and speedboat races, a fair, and a firework display. Thousands of people come to this annual event and the disturbance does not help the wildlife. This year was no exception, the festivities resulting in almost all the wild ducks deserting the reservoir. In previous years eggs of moorhens and coots have been seen floating in the water, presumably having been dislodged by the waves from the speedboats. The weekend usually results in the only pair of Great Crested Grebes abandoning their nesting attempt, but this year none tried at all. Sunday, 28 May, was a depressing day, what with the noise and disturbance. None of the House Martin boxes attracted occupants but, much worse than that, the young Tawny Owl which had recently hatched out in the wood was found dead. The birdwatcher who found it was attacked by the parent owl when he tried to pick it up and received a deep scrach on his neck from the owl's talons. The wound was not serious and the victim soon recovered after a nip of Glenlivet.

JUNE

June is usually a month with very little migration through the area, birds are busily feeding their broods and bird song tails off towards the end of the month. The first week of June is sometimes a period when the last Black Tern passes through and if the conditions are suitable a wader or two may linger on the muddy shore. This year the first week was one to remember; it all started when a strange warbler was heard in some Willows on the eastern marsh on 3rd. The song was loud and clear and could be heard across the reservoir which is only a few hundred yards wide at that point. Eventually the songster was traced to a group of willows by the marsh. It was a very active bird, continually flying from the top of one willow to another but during the relatively short periods of time when it was still a description was finally compiled. It was obviously a *Phylloscopus* warbler, a shade larger than a Chiffchaff, and it looked plumper, with a pronounced pale yellow eye stripe extending well behind the eye. The upper breast and throat were tinged yellow, fading to an off-white colour on the lower breast and belly. A most striking feature was the yellow undertail coverts. The upper parts were of a uniform fawn colour and at close range a yellowish but very faint wing bar was discernible, apparently due to light tips to the greater coverts. The legs were brown not black.

On the face of it the description sounded similar to the famous warblers of Drente in Holland. At least two of the warblers were observed singing in that locality in the summers of 1964 to 1967. They remained unidentified even though tape recordings of their songs have been heard by ornithologists from many countries in Europe. Fortunately we were able to obtain recordings of the Brent bird's song and later compared them with records of certain other Palearctic warblers. None seemed quite the same. The Greenish Warbler sounded similar but a recording of an Iberian Chiffchaff showed a closer resemblance. Eventually we found another recording of an Iberian Chiffchaff which was very similar. Sonagrams were made and compared with the birds of Drente and the Iberian Chiffchaff. The birds of Drente remain unidentified but the close similarity of the sonagrams of the Brent bird and the North African-Iberian Chiffchaff left no doubt as to its identity. The first record of this subspecies for Britain. The very faint wing bar confused the picture at first but then a few ordinary Chiffchaffs do have a suggestion of a wing bar, due to abrasion of the greater coverts rather than definite light tips to those feathers. Despite extensive searching the next day and the following week, nobody managed to locate it again.

This is the fifth and last year of the BTO Atlas scheme. Although most counties are carrying out theirs on a 10 km basis, the London Natural History Society decided to cover their area in more detail and adopted the tetrad as the basic unit. As there are twenty-five tetrads in a 10 km square the amount of work involved is considerably increased. Our area falls in part of six tetrads, so that each species may have to be proven to breed up to six times. Basically, the Atlas scheme involves mapping the breeding birds of the whole country on a presence or absence basis. There are three categories; possible breeding, in which a species has been recorded in the breeding season in a suitable nesting habitat, and may possibly be breeding in the area even though no further evidence is obtained. Probable breeding, when a bird is heard singing on more than one date in the same place, or where courtship is observed, or anxiety calls from adults suggesting probable presence of nest or young. The third category is confirmed breeding, such as an occupied nest, birds collecting food, carrying faecal sac, distraction display or recently fledged young.

June and July are the best times to prove breeding so that much of our activity in June was directed to making a special effort to prove breeding for those species we suspected as having bred, but which have eluded us for the last four seasons. We did quite well adding Redpoll and Goldcrest to the areas list of proved breeding birds. In addition, the Lesser Spotted Woodpecker probably bred. The Redpolls were found nesting on Barn Hill and were the first to be proved breeding in the area since the first decade of the present century. This recent recolonisation of the area reflects its huge population increase. Redpolls are now three times as common as they were in 1964 according to the BTO's Common Birds Census.

The Goldcrest was seen collecting food for its young in one of the old yews growing in the Saxon churchyard near the reservoir. This was also the first known breeding since the first decade of the present century in the area, whilst the Lesser Spotted Woodpecker had not been known to hold territory since the 1930s. In the end we recorded 50 species proved breeding, another four probably doing so, and six more possibly breeding.

In addition to the Atlas scheme we still had our census work to keep up. The census of the farm at Barn Hill was a major undertaking because of its size. Rising to 282 feet, Barn Hill is the highest point in the study area. James Edmund Harting writing in 1866 described it as a 'Woody height overlooking the parishes of Kingsbury, Willesden and Harrow'.

It is pleasing to relate that now over one hundred years later the same description is still applicable, at least to the top of the hill. Although much frequented by the public the dense scrubby areas on the Hill still provide nesting sites for a comparatively diverse and abundant avifauna. Adjacent is Big Bush Farm covering an area of nearly 250 acres. It was, until 1967, a dairy farm with moderately small fields and thick old mature hedgerows. In that year the farm changed tenants and became primarily an arable farm, but with a few fields kept aside for grazing and hay. On one side of the farm, which is bisected by the B4566, virtually all the existing hedgerows were removed although the mature timber consisting of oaks and elms was left.

The presence of these old trees on the farm and the fact that gardens and a thick woody border surround it on most sides are important features, enabling the farm to retain much of its bird life despite the removal of most of the hedges. The farm and the adjoining scrubland of Barn Hill have been censused as a unit, as many farms in the British Isles contain up to 15% scrub or copse. The thirty acres of scrub form about 10% of the total area. Census work shows that the farm supports a considerable range of species which together average just over 460 territories, which is about 394 per square kilometre. This compares favourably with rural farms which usually have between 100–400 pairs per square kilometre, so that the farm at Barn Hill ranks among the high density farms of the country. See diagram 4, page 21.

Surprisingly, only about 6% of the territories detected were based mainly in the surrounding gardens, but there was evidence that some species such as Blackbird and Starling were using gardens as feeding areas, and possibly this is one reason why the Blackbird density on this farm of 100 pairs per square kilometre ranks amongst the highest farmland densities in the country for this species. There is no doubt that this unit of open land enables many more species is survive than in an equivalent area of houses and gardens. In fact 36 species of birds have held territory there in the last few years. These included all the suburban garden species as well as Tree Pipit and Lesser

Spotted Woodpecker, which are not found on any other part of the study area. The pie diagram (page 21) shows the percentage of the total number of pairs each species accounts for in the farmland bird community.

During and after the Second World War a considerable part of the land adjacent to the Welsh Harp was given over to allotments, but these have been gradually reduced and now there are less than 40 acres left. In 1968 and 1969 we chose a plot of 29 acres to census. Our area was a mosaic of small holdings with a considerable variety of crops. Several hedges still remain containing some mature deciduous trees. This habitat is rich in bird life with a community structure resembling that of farmland, involving twenty-one species. (See diagram 5, page 22). The allotments are the main stronghold of the Linnet and Goldfinch in the area. Skylarks are also at their highest density here. The conversion of the war-time allotments into playing fields must have reduced the finch population of the area considerably, at least to the pre-war level.

JULY

In early July when many young birds have left the nest the first indications of the autumn migration become apparent. A visit to the reservoir in the hours after dawn often produces a wader or two. Redshank, Greenshank, Green Sandpiper or even Curlew may pass through. Small numbers of Lapwings continue to fly over the area heading for their post breeding quarters further west, and sometimes a flock of these beautiful waders may stop a while to rest and feed on the farm fields at Barn Hill.

A feature of this time of the year is the appearance of hundreds of Swifts over the reservoir of an evening. Small numbers are regular in June, but the much swollen population of late June and early July will, largely, be breeding adults from a wide radius, now gathering insects for their newly hatched young. It is a common sight to see over five hundred wheeling about in the sky, sometimes large concentrations can be seen away from the reservoir over the houses or wherever the aeroplankton is rich. Not only Swifts but Black-headed Gulls, Starlings and House Martins may be seen indulging in these mass aerial manoeuvres. Later when the vast urban population of ants choose the day for their mating flights the pavements become a seething mass of insects. Armies of them pour from every crack to form rising columns of insects, then even the sparrows exploit this temporarily abundant food source, displaying flight agility one would never have credited them with.

On cloudy and sultry evenings, especially when there is a touch of drizzle in the air, the Swifts feed much lower, so low in fact that they swarm on the insects at our head height. In these conditions they can be caught and ringed. The method used is simple; two people hold the end poles of a mist net which is kept horizontal just above the ground. When a Swift approaches the net is quickly flicked up. The timing has to be split second as Swifts are agile fliers and can avoid the net if it appears when they are more than a foot or two away.

Swifts are not one of the more pleasant birds to ring. Not only are they infested with up to a dozen or more blood sucking flat flies, but their claws are also needle sharp and can draw blood without difficulty. It is thought that the flightless flies cannot live for long on humans. However, one of our ringers found one walking around his leg in bed the next morning, and another had one crawl up into his white collar at a dinner party.

Although we have had no foreign recoveries from our Swifts ringed so far, one was caught by another ringer at Kempton Park in Middlesex six years after we had ringed the bird, and another was subsequently heard of at Tewinbury in Hertfordshire two years later. We have also caught a bird ringed at Barn Elms Reservoir in Surrey and retrapped some of our own birds caught in previous years. Ringing has shown that the Swift has a survival rate of about 80–85% which is much higher than most species of small birds. Some Swifts reach the age of 15 years. Only somewhere in the region of 50–60% of many small bird populations survive each year, and only one in a hundred Blackbirds, for example, lives ten years. We know of none over 12 years old.

Early July is a time when wildlife seems most abundant, many young birds have fledged, and the reservoir and the few ponds still remaining are teeming with invertebrate life. A few marsh frogs have recently appeared in the area and these call frequently from the edge of the reed beds in late summer. The common frog now appears to be extinct at the Brent Reservoir but only a decade ago frog spawn was abundant there. A mat of Amphibious Bistort grows in places on the edge of the reeds and these are favoured places for the Little Grebe to claim its territory with its loud and strange whinnying trill-like call which echoes across the reservoir. The reed bed itself is alive with the song of Reed and Sedge Warblers whilst the Reed Buntings increase their song output about this time. This peaceful atmosphere is acceptably broken by the frequent startled calls of Moorhen and Coot, but not so acceptably broken by the roar of passing trains or aeroplanes and the hornblowing of bad-tempered motorists.

It is just when nature seems at its peak and many of the marshland plants are in full flower that disaster often strikes—fire. A year never goes by without one or more large fires ravaging part of the area. Most of these are started by children, although some are accidents usually the result of lighted cigarette butts being discarded in the long grass. This year a fire was deliberately started in the *Phragmites* reed bed, half of which was destroyed before the fire was extinguished with the help of the local fire brigade. Soon after the flames were put out I saw a Reed Warbler, presumably a female, the feathers on her back badly singed; quite possibly she had stayed on her nest right to the last moment until the flames consumed her young family. One wonders whether such thoughtless violence against nature would happen so often if these children had been given an opportunity to learn about wild life at a local field centre and taught to respect it.

Towards the end of the breeding season in July a number of species form temporary roosts in the area, composed mainly of young birds. A large Starling roost has for several years been formed in late summer in the area of willow swamp at the eastern end of the reservoir. This roost, numbering several thousand birds, is mainly of juveniles some of which come from a nearby rubbish dump. A feature of this roost is that the slightest disturbance breaks it up, and by September it is considerably reduced, most of the birds having died or disappeared. Ringing these Starlings has shown that they disperse over a large area. Recoveries have been had from Fakenham in Norfolk ninety-eight miles to the northeast, Romford in Essex and Northfleet in Kent. Others come from Paddington and Kensel Rise, New Southgate, Elmers End and Catford. Only four out of fifteen recoveries have been of local birds.

The large thrush roosts in the area are mainly of juvenile Blackbirds and Song Thrushes. At this time of the year many of the adults have started their moult and stay away, in fact they do not start appearing again in any numbers until mid-September. Presumably the increased energy demands of moulting together with the frequent shortage of food at this time of the year, discourages them from making the journey to the roost. There is frequently a certain amount of activity and squabbling at the roost before the birds settle down, and this would be harmful for growing feathers, which still have a blood supply and are very easily damaged.

We have ringed about six and a half thousand Blackbirds at the main roost now, and these have produced just over 250 recoveries. In contrast to the pattern of recoveries for the Starling, all but four of the Blackbirds were in London and only one in eighteen had moved more than three miles. These results suggest that the population is a highly residential one.

At this time of year, also, large roosts of House Sparrows occur in shrubberies; one is formed every year on the end of the main Blackbird roost. The Blackbirds go to roost late, most arriving after sunset and some leave it so late that it is too dark to read one's own notebook. They also rise early and leave the roost well before sunrise. The sparrows are 'early to bed', usually just before, or just as, the sun sets and do not leave the roost until the sun rises. They have a good hour more rest than the Blackbirds.

Throughout the month evidence of autumn migration increases. There is no lull as there was in late March. Before many birds have finished rearing their young others are already returning to their post-breeding quarters. By the middle of the month moulting, surface feeding ducks arrive, species such as Garganey, Teal, Shoveler and Gadwall, and spend most of their time feeding in the shallows of the eastern marsh.

In some years the passage of waders is much better than in others. Although the amount of mud exposed is the most important factor in determining the numbers at the reservoir, years when the water level is low and ideal conditions prevail are sometimes poor. It seems likely that the volume of waders passing through England varies annually. If one is lucky the last week of July can be good for waders and an early morning visit may be quite exciting. Sometimes, on still mornings, the first mists of autumn may form over the reservoir and when these clear waders such as Common, Green and Wood Sandpipers, Dunlin, Redshank, Greenshank and Snipe may be spotted. Numbers are small, and regular birdwatching is needed before all these species can be seen. The eastern marsh is a difficult area for seeing waders as there are so many places where they can hide in the vegetation or behind the tin cans, petrol cans, railway sleepers and a whole host of miscellaneous articles which find their way into the reservoir and become stranded on the mud when the water level falls. Walking out on the marsh is not a good idea as this disturbs everything, and waders at the Brent tend to fly right away at the first disturbance as there are very few alternative places where they can settle.

AUGUST

During August, migration through the area gathers momentum; warblers, flycatchers, wagtails and hirundines are much in evidence. When mist-netting started in the area in 1962 we were amazed at the numbers of migrants caught. The increased attention paid to the scrubby sectors (admittedly at the expense of the reservoir itself) revealed far more warblers passing through the locality than had been imagined. Lesser Whitethroats, which we had previously thought of as rather a scarce bird, were turning up regularly in the nets. Garden Warblers too were prevalent; these are secretive species on migration, unlike the Chiffchaff which frequently sings in the autumn. In 1964, two Nightingales were caught and ringed, one in July and one in August. We notice that the numbers for certain species of warblers are much lower some autumns than others. Lesser Whitethroats were common in 1962, in other years we hardly caught any. There have been years when more Garden Warblers have been trapped than Blackcaps, although it is usually the other way round.

The reservoir is also interesting because the autumn flock of moulting ducks continues to grow in numbers, with Shoveler, Teal, Garganey, Mallard and

Gadwall as the main members. In recent years Gadwall have become more regular and the autumn flock has reached over thirty birds later in the year. They are usually to be found dabbling in the shallows at the eastern end of the reservoir, together with the odd heron and a few waders.

For part of August the water level was quite low but, despite this, the passage of waders was well down on most years when similar conditions prevailed. A few Greenshank, Ringed Plover and Dunlin, as well as Common Sandpipers and Redshank, were all that turned up. In the dry autumn of 1959 when 70 visits were made to the reservoir, an average of 9.4 waders was seen per visit. The next year, which was fairly wet with a high water level most of the time, produced an equivalent figure of only 3.8. This year the average is 5.6.

By mid-August virtually all the marsh flora is flowering and dragon flies are abundant, together with a host of butterflies. The patches of water mint, which is common on the marsh, are favourite gathering places for butterflies to feed. Small Tortoiseshells, Peacocks, Red Admirals, together with Essex, Small and Large Skippers, Small Coppers and Green Veined Whites can all be seen. Other species, too, such as the Small White, Large White, Wall and Meadow Browns, Holly and Common Blues add to the prevalent wildlife.

August is a good month for surprises and this year was no exception. The bird of the month was undoubtedly the Spotted Crake on 15th. This bird afforded brief but excellent views to the observer whilst it fed, skulkingly, along the edge of the *Phragmites* reed bed. This was the first record for the area since 1959 when one was seen on 23rd August. A Kingfisher on 20th was almost as rare; I have only seen two others here since the 1962–63 winter which exterminated them in many areas. This bird was a very welcome and attractive visitor to the area. It spent a lot of time fishing for Sticklebacks from branches of willows overhanging the reservoir edge, and was later joined by another bird. For a few days one of the Kingfishers took a liking for perching on the side of the bridge which has recently been constructed across the River Brent just before it enters the reservoir. The alterations to the river have resulted in a stretch of bank which is high and very steep. When vegetation has colonised the bank or been planted, it may be suitable as a nest site for Kingfishers. It is probably too much to hope for as none have nested in the area for over seventy years, ever since the river became badly polluted with industrial waste. The river, although classified as class two by the River Pollution Survey of 1971, has more the characteristics of class three. These are rivers with a fish population restricted in species, and receiving discharges of solids in suspension which have affected the river bed. In any event Kingfishers have kept away from the river in summer for many years. There is, however, no reason why the river cannot be improved to class one, except that it would cost money and, despite lip service to the contrary, most Authorities do not consider money spent on cleaning up the environment as money well spent.

On 27th August this year, whilst listening to the piping calls of a half dozen Common Sandpipers which were flitting between the debris on the mud and busily feeding on myriads of tiny flies which were just hatched, a loud and shrill 'ki-ki-ki' call of a Kestrel ripped the air. The sound came from behind me and on looking round I saw two male Kestrels engaged in fierce combat on the top ledge of the tall block of the British Oxygen Company offices adjacent to the reservoir. Both birds were screaming at each other, each with the claws of one foot firmly embedded in the other's body whilst kicking with the free foot. Wings were flapping and feathers flying, and so tightly were they interlocked that when one bird kicked the other especially hard both fell off the ledge, landing on the one below, a drop of some twelve feet. The fight continued, but now their bills were brought into action, ripping lumps of feathers from the opponent. Again, they fell to the next ledge, then again to the one below, seemingly oblivious of their hard landings. Eventually they reached ground level where they could no longer be seen. The screaming ceased and one bird flew off over the reservoir. For several seconds I thought it must have been a fight to the death but to my relief the second Kestrel appeared and flapped off in the opposite direction towards some railway sidings. These birds were probably males of two adjacent territories and found that their land overlapped. The boundary obviously had to be settled once and for all, there and then.

SEPTEMBER

September is perhaps our most exciting month of the year; migration is in full swing and anything can turn up. Large falls of migrants sometimes occur in the first two weeks of the month and a visit in the early morning is well worthwhile. These big arrivals usually occur after a night which has begun bright and clear but becomes overcast in the early hours of the morning, followed soon afterwards by heavy rain. If there should be an easterly wind blowing as well, a visit as soon as the rain stops is a must. We did not record any big falls this September because in common with recent years the autumn has been anticyclonic, with many clear nights. The last gigantic fall of birds recorded was on 6th September 1968. There had been really heavy rain the previous night and this kept up until the early evening. A visit to the wood after the sun came out revealed that something really extraordinary had happened. The Oak, Bramble and Hawthorn scrub was swarming with migrants too numerous to count, everywhere one looked it was the same. There were Willow Warblers, Chiffchaffs, Garden Warblers, Blackcaps, Whitethroats and Lesser Whitethroats, desperately trying to make use of the few hours of daylight left to put back some of the fat which they had used up on their migratory journey. Few of them would have had much success searching for food earlier that day with the rain storm still raging. The next night was fairly clear, but when we returned with mistnets the following evening it was obvious that very few birds had left the night before. We caught nearly 30 warblers that evening in a very restricted area where we usually put up mistnets to catch thrushes going to roost. There must have been hundreds in the whole wood and just how many in the open space around the reservoir is anybody's guess.

A strange phenomenon, which begins in August most years and continues throughout September, is the appearance of small numbers of escaped cage birds, usually in the reeds bordering the reservoir. The most frequent species involved is the Avadavat, sometimes called the Tiger Finch. This bird is common in India where it has a preference for damp areas. Its regular appearances around the Brent Reservoir in Autumn may be the result of a migratory instinct still present in these once captive birds, but where they come from and where they go to remains a mystery. Avadavats are small birds not more than four inches long and are seed eaters except when feeding their young. In common with many other species of finch they are insectivorous at this stage. Originating in a semi-tropical area there is little chance of their surviving severe weather unless they leave this country in winter. The best year for these birds was in 1967; in that year the first bird turned up on 10th August, this was eventually joined by eight others. Five of them were caught in a mistnet and put in a large aviary for their own safety—the last one survived another three years in captivity. The remaining four birds which avoided the mistnet were last seen on 6th October. During the rest of August and September that year, several other escaped birds made fleeting visits to the reservoir. These included a Budgerigar, a Java Sparrow and a Red-headed Bunting In other years several types of Waxbills have found their way to the reservoir, as well as a Bulbul and a Napoleon Weaver. Other escaped birds have been seen but remain unidentified. Eric Simms on the adjacent Dollis Hill can add Rosy-faced Lovebird and African Grey Parrot to the list.

September is perhaps the most likely month to see Shelduck on the reservoir although only one was reported this year. However, five early Wigeon on 2nd and a Red-crested Pochard from 17–19th made up for this. The origin of the Red-crested Pochards which turn up from time to time in the autumn will never be known for certain. We do know that Abberton Reservoir in Essex has a build-up of continental birds in the autumn, so it is conceivable that our birds have the same origin. Other distinguished visitors this year were four Little Gulls on the first day of the month, all previous records were of single birds. We heard later that the autumn passage of Little Gulls was heavier than usual, and at Dungeness in Kent there was the unprecedented number of 106 on 9th September.

Pride of place must go to the Purple Heron on 12th, the first for the area. It was seen on a Tuesday morning flying lazily out of a patch of reed grass and heading towards the eastern marsh where it was later seen feeding in the shallow water. Unfortunately, its visit was brief, and despite extensive searches the next day the bird could not be found. Last September the birds of the month were Arctic Skua and Red-backed Shrike. The former spent several hours at the reservoir chasing Common Gulls in true Skua fashion, attempting to make them disgorge their last meal. It had little success however and settled on the water to preen before flying away.

An interesting feature of the last few Septembers has been the late departure of Swifts. It used to be unusual to see a Swift in late August but last year there were no less than eighty still present on 28th August and 12 remained until 4th September. Similar numbers occurred again this year, and apparently a number of nests still contained young at the beginning of September in other parts of the country. Perhaps the run of cold damp springs we have been experiencing delayed the onset of the breeding season.

Towards the end of August, and throughout September and October, a build-up of the Little Grebe population occurs. Although the numbers involved are now only a fraction of what they were in the 1950s when up to 121 together were seen, the population still regularly tops thirty. They are difficult to count because of their busy diving, and when this activity is not taking place the birds rest in the shallows of the eastern marsh and among the reeds bordering other parts of the reservoir, where they can hide out of sight. Although we have little evidence one way or another, regular counts suggest that the reservoir is a gathering ground for Little Grebe after the breeding season for moulting purposes, and that they are not part of a daily passage of different birds going through.

September was marred this year by the news that, despite public opinion against it, proposals by the British Waterways to construct six-storey tower blocks of luxury flats, on a playing field adjacent to the eastern marsh, was approved in principle by the Department of the Environment. Clearly, the pressure to encroach on this vital open space in the heart of suburban London is mounting. A stand has to be made, not only from the ornithologists, but from all groups who use the area for leisure activities of one sort or another. Otherwise this vital open space will become substantially choked by urbanisation, reducing the quality of life in the area a little closer to that 'enjoyed' by battery hens.

OCTOBER

October is the month when the bird life of the area changes dramatically once more. All but a few summer visitors have left and the winter visitors are beginning to arrive. I was away for the last week of September and the first in October, and the difference in bird life on my return was very striking. The only summer visitor left was a Chiffchaff, but Redwings and Fieldfares were everywhere, the former especially on the playing fields where groups were feeding along with Blackbirds and Song Thrushes. Small flocks of Meadow Pipits fed on the grassy shores of the reservoir whilst a few Redpolls and Goldcrests searched for food in the willows. Several Snipe were flighting over the marsh and the first Jack Snipe had already arrived.

Perhaps the most outstanding feature of this October was the enormous passage of Redwings at night, at times one could hear three or four calls a minute; thousands must have passed over the area and this local passage was reflected throughout the country. It is thought that anticyclonic conditions over Scandinavia, with easterly winds, drifted many Redwings which would have wintered in France, over to England.

By 22nd October the Coot population had risen to 240, the largest flock for a good many years. Also on that date our wintering Water Rails first made their return noticed with their piglike squealing. They were in the reed mace on the eastern marsh and ventured out cautiously for brief spells to feed. A quiet approach is needed to see these birds, and plenty of spare time to sit hidden in the reeds, as they will only come out into the open when they think no one is about. Any sudden noise or movement, a starting gun for a sailing race, an aeroplane or even a gull flying over, will send them scurrying back into the dense marshy vegetation.

The last of the summer visitors, the Chiffchaff, lingered no later than 14th October although a House Martin was reported over the Blackbird roost on the 21st. In previous years Swallows have sometimes continued to pass through the area right up to the 28th. We have never managed to see a November Swallow, although they are not uncommon on the south coast at that time. It is difficult to find the latest date for Chiffchaff and Blackcap as both these species have been known to winter in the area on occasion. We have had four species of tern in October. In 1957 a Black Tern was seen as late as 13th October whilst in 1967 an Arctic Tern lingered on until 17th October. The Wheatear, which is usually our first spring arrival, also features amongst the latest to leave, one bird in 1959 actually remained until 1st November. Perhaps the most outstanding late date was the hen or immature Yellow Wagtail on 26th November 1960. This bird may have spent the winter in the London area as other birds were seen in December at Petts Wood in Kent and two birds wintered at Beddington Sewage Farm in Surrey that year. One had done so in the previous winter and these were the first and so far the only records of the species in the London area for that season. They usually leave the country in the autumn for southern Spain and North and tropical West Africa.

One of the characteristics of early October is the sudden appearance of large numbers of Blackbirds feeding in the hawthorn bushes and willows adjoining the reservoir, or on the numerous playing fields nearby. We used to think these birds were migrants, perhaps, from the Continent as many thousands of Blackbirds do pour into England about this time. However, since our ringing activities began at the Blackbird roosts many of these feeding Blackbirds bear a ring on one of their legs, and although some migrants probably do stay a short while, we think the majority are local birds which have finished the moult which kept them unobtrusive in August and September.

We have been studying one of the Blackbird roosts with the aim of learning more about the functions of communal roosts. It is in use throughout the year and it seems likely that this particular roost has protection from predators as a prime function, but other benefits may exist at different times of the year. There is no doubt that the Blackbirds pick the most sheltered and densest bushes, and this has been tested by comparing the wind speed in the bush and outside on two sensitive anemometers. The greatest reduction of wind speed occurs in the bushes chosen for roost sites, and the reduction may be as much as two-thirds. The nightly temperatures inside the bushes are however no different from those in more exposed positions. As the chilling effect of the wind is the most important factor in heat loss we tried to quantify the effects of the wind reduction on the metabolism of roosting Blackbirds.

We roosted birds singly overnight in their usual roost sites and in more exposed positions in cages made of wood and net. Birds in the traditional sites only lost about 60% of the weight lost by birds in more exposed situations on windy nights when the temperature was around freezing point. There was only a saving of some 15% during calm nights. The strange thing about these Blackbirds is that they roost several feet apart so that they cannot warm each other as Starlings probably do when they roost shoulder to shoulder.

On some other nights during the winter we roosted different numbers of Blackbirds in large cages suspended in the roost. When there were two birds per 10 cubic feet the average weight loss per hour was in the region of 0.38 gram, but when there were six birds in the same space the loss was 0.54 gram. At a density of one bird per cubic foot nearly 0.7 gram was lost each hour. Blackbirds will not tolerate each other too close and squabble if a minimum distance is not observed. If four birds were put in one of the large cages they would always retreat to the four corners. Blackbirds are not sociable birds like Starlings, and as they seem seldom in danger of dying of starvation in London the desire to conserve more heat by mutual warming presumably does not exist.

By being dispersed, birds are more difficult for predators to locate than they would be clumped together. However by roosting in the same general area there will always be a few birds awake at any one moment, so that it would be difficult for a predator to sneak up unseen. It has been suggested that an alarm call given by an unthreatened bird when it spots a predator may momentarily distract the predator from its prey, enabling the latter to escape.

We do not think the theory, which suggests that birds roost communally to learn the whereabouts of the latest food source from each other, applies to suburban Blackbirds as there is a super-abundance of food, even in the cold spells of weather with snow cover, and food is so dispersed that there would be little advantage in learning where one supply is.

We have arrived at the conclusion that the prime function of this particular roost is protection from predators. The most important predator in the area is the cat, hundreds of which roam about at night in the gardens where there are relatively few safe roosting sites. The vegetation chosen for the communal roost is away from the gardens and is the densest in the area. The way Blackbirds roost also seems predator orientated for reasons mentioned above. The only predator which attacks the roost regularly is the Tawny Owl, this is sometimes mobbed furiously by several Blackbirds if it enters the roost at dusk. This anti-predator behaviour is not fool proof, and analysis of local Tawny Owl pellets has revealed a number of rings from Blackbirds, neatly tucked inside.

NOVEMBER

The last few days of October and the first week or so of November often coincide with the peak of the autumn influx of thrushes and finches, particularly if the wind is from the east. The most spectacular movement for many years took place on 1st November 1959. I arrived just after sunrise and one of the first birds seen was a Rock Pipit feeding on the exposed stony shore at the western end of the reservoir, it flew off further up the bank with its characteristic harsh *tsip* call to join a little group of Meadow Pipits busily feeding on some small flies which were crawling between the stones. Early November is the most likely time to see this uncommon visitor to the area. For the first few hours after dawn there was some overhead movement that morning, but nothing unusual for the time of the year, then suddenly at about 10 a.m. the first large flocks appeared; they were all Skylarks, some three hundred passing over in the next hour, all heading west. The movement was on a broad front and it was impossible to count more than a small section of about half a mile wide. By 11 a.m. Fieldfares were appearing and in the next two hours over a thousand birds flew over, together with hundreds more Skylarks and several large flocks of Chaffinches and Linnets. Observers from other parts of London also recorded these movements and, from an average count of nearly 500 Chaffinches an hour, it has been calculated that 60,000 birds of this species passed through the northern radius of the London area on this day alone. Most of the migrants flew over without stopping, but small groups of Fieldfares dropped into the hawthorns with their loud 'chacking' calls, to feed briefly on the berries.

When movements like these happen there always seem to be a few rarities caught up with them. On this occasion the Rock Pipit, a very late Wheatear and a Red-legged Partridge were all unusual for the area. On the other hand, there does not have to be a movement to produce the unexpected. Snow Buntings, a rare visitor to London, are most likely to be seen this month but only after a day or two of easterly winds. The only record of an Osprey at the reservoir was in November 1961. It turned up on a Sunday afternoon, of all times, when there were some sixty boats out sailing. The bird came in from the south-west and did a terrific slanting dive but did not enter the water. It then glided and flapped around, putting up all the gulls as it hovered looking for fish, before flying off.

The winter season is well underway and by mid-November most of the Elms and Oaks have usually lost their leaves. This year the trees retained their leaves rather later, possibly because of the absence of strong winds, and some fine displays of autumnal hues were to be seen. It is a sad thought that many of our mature elms will never bear leaves again—the Dutch Elm disease is rampant in the area and many trees have been condemned, marked with large white crosses. A whole row is to be cut down in our ringing site and many more will follow. There will be profound changes in the avifauna of the area as a result, especially with hole-nesting species unless we can put up a large number of nestboxes over the next year or two.

We had some good news in November with confirmation that the Field Centre near Birchen Grove would go ahead and a special meeting was held on the 12th to discuss the final plans. Several Brent Councillors and the Directors of Education and Planning were present. The result was a satisfactory arrangement of compatible amenities for the area. The site discussed was some 50 acres, of which 26 acres will be public open space adjacent to the reservoir. Nearly 13 acres are to be set aside for a Nature Study area for schools and interested members of the public. Bordering this site will be three acres of allotments, over five acres for a central parks nursery, and two acres for an adventure playground. The Field Centre is therefore buffered by gardens, allotments, a plants nursery, and an adventure playground where the existing vegetation is to be left.

The Field Centre is likely to receive something in the order of 30,000 children a year and will undoubtedly degenerate under human pressure if a carefully controlled management plan is not carried out. We also want to create new habitats and modify existing ones in the field centre area for educational purposes. We had in mind an acid bog with insectivorous plants, a meadow, closed and open woodland, scrub, ungrazed grassland, stony ground, neutral or alkaline marsh surrounding a pond, and a stream; one already runs along one side of the area but is heavily polluted and will need cleaning. It will be interesting to see how the breeding bird carrying capacity of the area is affected by these changes and it is essential that we carry on with the mapping census each year. It is an exciting new venture and one which we hope will have an important part to play in teaching the children of Brent to respect their environment and the life which dwells there.

The situation at the other end of the reservoir is less promising and the news of the proposed development adjacent to the eastern marsh has upset many people. A meeting was held recently involving some Barnet councillors and other interested people, to discuss what might be done to stop this sort of exploitation. It was unanimously felt that there was a strong case for the formation of a pressure group, and so the Welsh Harp Conservation Group was formed, its aims being to protect the Welsh Harp area from any further encroachment by industrial and residential development, and to improve the area for recreational purposes. The group will press for action to clean the polluted waters of the reservoir, and the streams and rivers which flow into it; the termination of the rubbish dump next to the reservoir; an extensive tree-planting programme to start in 1973, the Year of the Tree, and to continue indefinitely; the declaration of the eastern end of the reservoir as a wildlife reserve, and the establishment of a nature trail around the reservoir with a few hides overlooking the eastern marsh.

One of our first tasks is the organisation of a massive petition to help Brent Council in persuading the Department of the Environment to reverse their decision to allow further building on the banks of the reservoir. The Group aims to help co-ordinate the views of, and encourage discussion between, all users of the Welsh Harp area, the ultimate aim being to improve the area for public enjoyment and to cater for as many compatible amenities as possible.

DECEMBER

Having been away for two weeks my first visit was not until 10th December. It is another mild winter so far and the weather being so bright I decided to choose the day to photograph the two female Scaup which arrived on 26th November. Scaup on inland reservoirs are often viewed with suspicion because of the possible confusion with hybrid Pochard × Tufted Duck or Scaup × Tufted Duck. In addition, young female Scaup sometimes lack the white blaze on the forehead altogether, or else it is diffused and reduced in size compared to the adult. In these circumstances they can be confused with female Tufted Duck which often have white patches on the forehead. The only safe way is to check the bill tip, which is black on the nail only. The bills of Tufted Duck have a larger black tip. We managed to view the two birds through a telescope and could see the bill pattern resembled that of genuine Scaup. They constitute the thirty-sixth record of this maritime duck at the reservoir, all but four of these have been since 1946.

So far there are very few ducks at the reservoir for the time of the year and no Smew. In recent years Smew, for which the Brent is famous, have not turned up until Christmas-time, probably owing to the mild winter conditions. There was still plenty of interest; the Snipe numbers are building up, odd individuals being frequently flushed from the marshy vegetation near the water's edge. I had four exposures left on the film and wanted to use them up before making tracks for home. I did not have to look far. Suddenly, the air was filled with a metallic ching-like call and, looking up, I saw three Bearded Tits descending from high out of the sky to drop into the reeds just a few yards in front of me. There was a willow bush behind them and, in the hope they might fly onto it when I approached, I focused and took a light reading on the bush. As soon as they landed in the reeds they were silent and could easily have been passed by unnoticed as I was able to approach to within eight feet before the three distinguished visitors flew up from the reeds and landed exactly where I had hoped they would. I managed to use up all the frames before they, two females and a male, started to call anxiously and took off rising high in the sky with their typical wiring helicopter-like flight and headed in the direction of the eastern marsh.

The weather remained mild throughout the month with little wind and plenty of fog. The Blackbird roost seems to hold less birds this year, but for several years the proportion of young birds in the autumn has been lower than in the mid-1960s, suggesting poorer nesting success, and this may reflect a declining population. There is evidence to suggest that Blackbird populations fluctuate over a period of two decades, with peaks and troughs every eight or nine years. The population was rising throughout the late 1960s and if it follows the pattern after the previous peak in the mid-1950s one can expect it to decline for a number of years to come.

For a long while we were curious to know what happened at the roost in really dense fog. When visibility is down to 50–60 yards there is some evidence of a fall in numbers entering the roost. During the evening of 13 December 1970 we found the answer; the fog suddenly became thick, so thick in fact that one could not see one end of a sixty foot mistnet from the other. At times visibility was down to as little as 40 feet. Under these conditions there was a very considerable reduction in the numbers of birds entering the roost, which was almost silent. Several Blackbirds were found roosting in bushes not normally used; they were easy to approach and very reluctant to fly. Only a few Blackbirds were caught when the fog was at its densest, although Starlings were heard descending into the roost, but these were probably flying above the fog. Likewise, a House Sparrow which suddenly fluttered straight down to the grass and stayed there. Hundreds of Blackbirds must have been forced to roost in the gardens that night because even when the fog cleared a bit it was too dark to tempt more than a few in. One wonders how the birds managed in the days of the London Smogs.

The last two weeks of December were characterised this year by large numbers of diving ducks at the reservoir. By 23rd December there were no less than 375 Tufted Duck, 47 Pochard and one Goldeneye. This is the largest flock of Tufted Ducks recorded in the area, although on 11th December 1949 there were 505 Pochard and some 70 Tufted Ducks present. The weather has been too mild to bring unusual numbers down south, possibly the large flock which has built up over the Christmas period, when sailing activities were much reduced, contains birds displaced from Queen Mary Reservoir or Island Barn Reservoir. These two Metropolitan Water Board reservoirs have recently had sailing clubs established on them. Other reservoirs may follow suit, and it is likely that this will have profound effects on the waterfowl in London. It is hoped that a professional study will soon be started to look into such effects and to identify areas of conflict between sailing and birdwatching interests, in order to encourage intelligent amenity use of the other London reservoirs with as little disturbance to wildlife as possible.

The first Smew did not arrive until 22nd December, when a pair were seen diving in the shallows at the east end of the reservoir. The numbers did not top four before the end of the year and this does not look too promising for the rest of the winter. If reasonable numbers are going to turn up one should have about 20, at least, by Christmas. Presumably the mild winter is not encouraging the winter populations off the coasts of Sweden and Holland to move further south.

We have only had one Heron this month, gone are the days of the mid-1950s when one could count up to 25 at the reservoir in the winter. A favourite site was the wall of the dam to the reservoir; here the Herons would roost at times, each spaced about eight or ten feet apart. When the maximum number was present the line would extend nearly 100 yards along the wall. It was no unusual sight to see Herons parading up and down a stretch of bank, stealthily paddling in the shallow water for fish. One bird was seen to fly along low over the surface of the reservoir and actually land on the water and swim. In the days when Herons were more frequent it was also known for a passing Heron to drop into a nearby garden with a goldfish pond and raid that. There does not seem to be any logical reason why Herons should be so rare here now, as the population of the country has recovered from the crash in the 1962–63 winter. The reservoir is no more polluted than it was in the 1950s, but perhaps the increased disturbance and activity in the area are the cause.

We are at the end of another year, at a time when so many changes are about to take place that one feels rather doubtful as to how it will all turn out. With the Welsh Harp Conservation Group becoming more and more active there is hope. The latest resistance against the high density developments by the British Waterways Board involves a petition which looks as though it will receive several thousand signatures, and we have just heard that Barnet Council have turned down the second application, and the Waterways Board are to appeal again. If that organisation has its way there will be the probability that other private land will be sold for development and the process of urbanisation will choke the area eventually. High density tower-block developments are biological deserts. Observations were made in the Chalkhill Estate near the reservoir containing over 370 people to the hectare. There were no gardens, merely lawns and a few trees, and the breeding avifauna was restricted to seven species, Woodpigeon, Blackbird, Blue Tit, House Sparrow, Starling and Carrion Crow as well as Feral Pigeon, and most of these were dependent on one or two mature trees left standing. It would seem likely that the suburbia of the future will be far less suitable for birds than today. Nor will it be more suitable for people, unless conservationists can make the planners and the government see sense.

By the time you read this many changes will have taken place. Why not visit the area and see them for yourself, you may be pleasantly surprised—on the other hand you may not.

WOODLAND

by JIM FLEGG

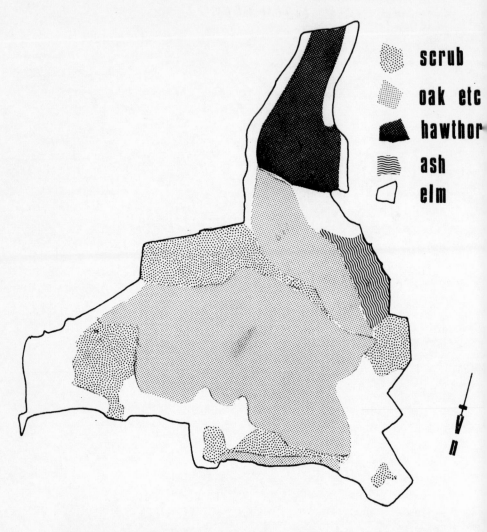

Sketch map of Northward Hill showing the dominant trees. The scrub areas contain some isolated hawthorn bushes, but are primarily grasses, bracken, bramble and other herbaceous plants.

Introduction

One of the few expanses of woodland visible on the south side of the Thames to the voyager sailing up to London docks is Northward Hill, lying draped over the ridge of London clay forming one edge of the Thames basin. Although rising to less than a couple of hundred feet above sea level, the wide-open view to the north towards the river, overlooking a belt of flat fresh marsh and arable farmland gives 'the Norrard' something of a commanding aspect. The view is a dynamic one, for not only is there shipping to watch, moving with the tide, but always there are birds to be seen—within the wood and over farmland, marsh and river alike. And beneath our feet in the wood lies one of the richest of soils for the fossil remains of our earliest birds: even bits of Woolly Rhinoceros are not irregularly turned up in farming operations, and not far upriver to the west is the home of Swanscombe man.

The wood is about 130 acres in extent, and is, as the sketch map shows, roughly triangular. The north-facing edge borders the marsh, and behind the boundary fence and ditch the land begins to rise, fairly uniformly and quite steeply, to a plateau of arable farmland (mostly cereals, potatoes and brassicas) along the southeast facing boundary. Here the land slopes gently away to the wide mudflats and extensive saltings of the Medway estuary. The southwesterly border skirts an irregular valley cutting into the ridge, full mostly of orchards and thus picturesque through most of spring and early summer, and attractively named Buck Hole. At its most southerly edge, Northward Hill touches the village of High Halstow. Here a blanket of modern housing, mostly for oil refinery personnel, swamps much of the old village apart from the attractive church and its neighbouring pub, the Red Dog. Discordant this housing development may be—and the cancer continues to spread—but for the small birds of the wood it does have its value, especially in winter.

The sketch map shows the most obvious vegetational characteristics of the wood. Probably around the turn of the century, Northward Hill was grazing farmland, and within the wood field boundaries can still be seen—such as huge gnarled hedge maples bearing the clear signs of having developed from laid hedges. Probably, too, most of the elms result from the natural expansion of old hedgerows, for the elm suckers very efficiently exploit new ground.

Oak woodland occupies about 44% of the total area—a very rough figure as there are many clearings within the boundaries. Many of the trees of the north facing slope are presentable standards, perhaps 40 feet high, but on the crest of the ridge some coppicing was carried out during the first World War, and the stools were left to develop untouched afterwards. The resulting timber is not specially beautiful in an aesthetic sense, and could be described as a forester's nightmare, but plenty of rotten branches, holes, and fallen trees make this a rich bird habitat. The shrub layer under the oaks is composed of the occasional spindly hawthorn or elder, and apart from dog's mercury, yellow archangel and bluebells in places, ground cover is deep leaf litter. Thus it is largely because of the insects living on the oaks that these areas provide the most suitable habitat for Blue and Great Tits, and here the majority of the nestboxes for the study population are situated.

Around the time of the second World War, and in the 1950s, the oaks supported the major part of the heronry, currently the biggest in Britain, for which the wood is renowned. Otherwise, though, the pure patches of standard oaks are not specially bird-rich except at their margins. Fortunately, High Halstow is a jigsaw puzzle of oaks and small bluebell-lined clearings, later filled head-high with bracken. On the margins and in the clearings, the absence of the dense canopy allows a much greater scrub and ground layer to develop (although with few new plant species other than blackthorn and bramble) and this 'edge-effect' produces an extensive fringe suitable for a much wider range of birds like the thrushes, finches and warblers.

The various belts of elm at their peak occupied about 26% of the wood, but as will become apparent, this figure is being much modified by the effects of Dutch Elm disease. Generally the undergrowth is composed almost entirely of elm suckers, which in places are almost impenetrable—a great place for emerging covered in cobwebs, twigs and caterpillers. Ground cover, other than occasional patches of unconquerable nettles or dog's mercury, seems to have given up the unequal struggle with the elm suckers. Even in its healthy heyday, elm is notorious for rotten branches, and such a situation is ideal for the larger hole-nesting birds like the Starling, Jackdaw and Stock Dove, which are capable of quick flights out into nearby marshes or farmland to collect food for their young. For a period, at least until the diseased trees fall or are felled, the situation for rotten-wood feeders and hole-nesters is going to be most satisfactory—but thereafter will come hard times as new growth replaces old.

Presently much the greater proportion of the heronry is in one of the belts of really tall mature elms, mostly over 60 feet high and some considerably taller with boles 3 or 4 feet across. Over the last four years, Dutch Elm disease has claimed every one of these trees, casting some doubt on the old folklore that decrees that the herons will leave before the trees die. A while ago there was a small colonisation of some robust oaks, but later a withdrawal to the taller elms took place. Presumably it is a combination of the 'open topped' nature of elms, and their presence as the tallest trees that caused the shift from oaks in the first place. We shall follow the changes in the heronry as trees begin to collapse, with interest and not a little concern, for it is by no means certain that the oaks will again prove acceptable. If not—no live elms exist for many miles around now, and with the nearest really tall trees beeches about 10 miles away on the downs (where the heronry originated before moving, for no certain reason, to High Halstow) the likely future for these 160 pairs of herons is problematic. A spur of rather younger elms close by the heronry supports a small but noisy rookery. The two species make uneasy neighbours and, especially early in the season, spectacular aerial conflicts are not uncommon.

The sycamore and ash areas—both relatively small—have a rather thinner canopy than either oak or elm, and consequently undergrowth is greater. Sycamore is generally regarded as a poor habitat for birds—probably justifiably so as it offers little in the way of food until late in the summer, when young tits and migrant warblers feast on the numerous aphids. In winter, however, the shrub layer brambles give shelter to large numbers of finches, Chaffinches especially, using the wood as an overnight roost. To the general annoyance of neighbouring farmers, hundreds—and sometimes thousands—of Woodpigeons also roost here.

The ash presents a different picture: coppiced over fifty years ago, it is in parts over-mature. Broken-off branches and holes are common (some natural, but many others excavated by woodpeckers) and here may flourish large numbers of breeding Starlings, with some Tree Sparrows, Jackdaws and Stock Doves. There is a puzzle here. It is difficult to reconcile the Great Spotted Woodpecker population since the war (usually about five pairs) with the many hundreds of woodpecker-excavated holes that are evident: it may be that pre-war the population was much higher but there is no supporting evidence for this theory.

Scattered throughout the wood are several areas of bushy thicket, varying in extent from a few square yards to about 2 acres. These thicket regions are usually composed of blackthorn or hawthorn. Blackthorn scrub is generally in smaller, very dense and prickly patches, almost inevitably impenetrable to humans and other large animals. Commonly these patches are used for winter roosts by finch (Goldfinch, Linnet, Redpoll) flocks and the overspill House Sparrow population that spends its days on the arable fields. The patches are so dense as to exclude ground cover plants, but along their margins taller herbaceous plants like thistles and teasels can develop, forming a useful autumn and winter food source especially for Goldfinches.

The hawthorn thickets are more variable in size and in density: many are more like collections of small trees with a good scattering of ground layer plants and bramble and rose. On the other hand, in the most southerly corner of the wood, the hawthorn growth is so dense that it could almost be described as climax vegetation. The canopy is such that abolutely no ground cover exists, and no timber trees have yet been able to establish a foothold. Here at dusk or after may be seen foxes and badgers, but to reach them the observer must penetrate the thicket, often using their tracks and normally on hands and knees!

Within the thickets, the branches are less crowded than in blackthorn, and this is reflected in the larger size of the birds using them for roosting—thrushes and starlings, and in a couple of special areas, Long-eared Owls too.

Foodwise, the blackthorn, despite a superb spring profusion of white blossom rarely produces a heavy crop of sloes, and indeed in most years it is difficult to collect enough for more than a couple of bottles of very delectable sloe gin. The hawthorn berry crop, depending on its size, may provide a substantial part of the food requirements of the thrush population—especially for the newly-arrived autumn migrant Fieldfares and Redwings. In years when the crop is poor, the numbers of thrushes in the wood during the day drop markedly once the few berries have been eaten.

These thickets, in the breeding season, provide safe nesting sites for a great many small birds of perhaps 20 species—most are not restricted to them but the Lesser Whitethroat and the Nightingale are. It is through these areas, and the open areas, that the breeding season predators like Jays and Magpies hunt, and here Cuckoos are most often seen looking for nests to parasitise, inevitably with an attendant train of scolding small birds.

Finally, in this introductory tour, we come to the open scrub areas. There are two major and two minor ones totalling 17% of the wood, and a host of tiny clearings better considered as features of the habitat in which they occur. The two major areas differ in character—a difference associated with the underlying soil type. That on the clay of the north-facing slope is composed largely of huge clumps of blackberry, with occasional hawthorns, rose and gorse bushes. There is a substantial area of bracken, and some stretches of rough grass with occasional stands of comfrey, teasels and a variety of thistles. The other area, running along the flat crest of the ridge, is similar at its eastern end, but changes markedly toward the west. Here there is about an acre of short, rabbit-cropped turf, with century and yellow-wort, on a thin skim of soil overlying sand, and surrounded by gorse. Occasionally in the past Woodlarks have bred, and for those with a taste for industrial landscapes, the sunset view, looking over the marshes to Gravesend and Tilbury docks in the distance, takes a lot of beating. Even to one so often in the wood, this view never fails to impress how close to the modern environment these 130 acres are, how precious they must be to the wildlife dependent on them, and in these circumstances how rigorous must be our measures to defend them.

URBAN AREA 1 Part of the eastern marsh at Brent Reservoir, showing developing silt beds and colonising vegetation—note the proximity of the houses. One of the largest concentrations of Smew in Britain over-winters here.

WOODLAND   2   Aerial view of Northwood Hill wood, High Halstow, showing the location of the woodland in relation to the village and surrounding farmland. For strictly woodland birds Northwood Hill is relatively isolated, especially in the north where the River Thames and the Essex coast lie just to the right of picture.

Photo: Aerofilms Ltd.

URBAN AREA 3 A further view of the eastern marsh, showing the position of nesting rafts, and some of the sailing dinghies which use the reservoir. Phragmites reed bed in the foreground.

4 Farmland at Barn Hill, looking west from the foot of the hill, with the A4140 road bisecting the area, and residential development in the background.

JANUARY

Perhaps it is best to start a narrative account of a birdwatching year in North-ward Hill in January, for at this time of year the wood may be most vital to the survival of some of its birds. The real bite of a northeasterly exposure is abundantly apparent as we walk the wood, and the short, often grey, January days give an unusual severity to the outlook. Underfoot the heavy soil has retained much of the winter's rain and progress on the slopes is often 'two forward and one back'. Gumboot weather!

More than any other feature, the shortness of the day imposes the greatest threats: not only is adequate food vital for a bird's normal functions and health, but in winter most birds have to maintain an insulating 'jacket' of body fat under the skin, together with reserves of fat to tide them over any periods of unusually severe weather. Add to this the fact that the available food supply, produced during the summer, has already been exploited for some months by adult birds and their youngsters of last year, making each mouthful the more difficult to find and the more vulnerable to competition.

Thus from the late daybreak right through till dusk, the majority of the birds of the wood will be feeding actively—some, like the Dunnocks and tits, remaining within its confines, others needing to make forays to the surrounding farmland or, in times of greatest stress, to the gardens of homes in the nearby village. In such circumstances it is important for the birds that they lose as little heat as possible—and hence the winter Robin, fluffed out to retain an extra-thick string-vest layer of insulating air close to its body. Energy requirements—drawing on the precious stores of body fat—are not much reduced overnight and may even be increased by the cold. The winter night is a long one, and temperatures may be considerably lower than during the day: thus the best available shelter is sought out for roosting, and squabbles over favoured perches may be just as violent, and just as important, as those taking place over a worm or a crab-apple fragment during the day. Exposure to a slightly colder draught of air right through the night may make a deal of difference.

So the hawthorn, the blackthorn, the bramble and the gorse serve their purpose—in late afternoon, small groups of birds begin to gather, twittering and chattering. The finches and thrushes usually move straight into the roost, but Starlings may gather at a series of pre-roost spots, before taking off to the main roosting area. The take-off moment at High Halstow (where often tens of thousands and sometimes hundreds of thousands of Starlings are involved) is preceded by a sudden silence—almost painfully oppressive in comparison with the previous racket—before the roar of wings hits you.

January is the month of the Long-eared Owls. With so many small birds roosting in the wood, food is not so difficult to come by after dark for them at least. With a little luck one or two of the group that spend the winter in the wood will be out hunting before the light has failed completely, and the astonishing silence of an owl's flight (due to a specialised feather design and the velvety feather surface) can be appreciated. During the day, the owls roost, more or less communally as is their custom, in some dense hawthorn scrub. A cautious approach—difficult enough in all conscience in a hawthorn thicket—will reveal the owls perched close to the main trunks. If our approach has been a quiet enough one, the owls will be 'at ease'—fluffed up and looking like Tawny Owls—and even in this 'plump' condition they are exceedingly easy to overlook. One small sound, any indication that they have been seen, and they freeze, first closing their feathers and slimming right down to the thickness of neighbouring branches. In this tight-feathered condition, sitting bolt upright, their cryptic colouration—a mixture of black, white and rich browns, serves them admirably for camouflage, and often the first that you know of their presence is a silent, heart-stopping, departure from right beside you. When they do fly, their agility is amazing: they are quite big birds, with a wingspan approaching 3 feet, but they can twist through the branches and twigs of hawthorn scrub without a sound—apparently just as easily as a Woodcock.

Late in the afternoon they can be quite eerie birds to watch: other than in the breeding season, they make little noise, and as you pass quietly through the roosting area, your movements are followed by startlingly fierce orange eyes. Only the owl's head moves (it can turn through more than 360°)—and the structure of the head epitomises the bird. The long feather tufts that give the bird its name have, of course, no hearing function, but the ears themselves are most remarkable: they are very large and also asymmetrical, a system that allows detection of very slight sounds and a very precise location of their source. Not surprisingly, experiments have shown the ability of Long-eared Owls to hunt successfully in total darkness.

Sometimes two or more owls will work co-operatively in hunting a finch roost: I've seen one bird flying fairly conspicuously down one side of a line of bushes, while another lurks quietly, Sparrowhawk-like, before making a dash at some of the birds disturbed by its fellow. Looking at the pellets (castings of fur, feathers and bone) that the owls regurgitate after a meal, tells us much about their food requirements: ironically, as bird rings also are indigestible, it tells us in our wood something of the fate of the birds that we ring. At High Halstow, and perhaps elsewhere, birds form quite a large element of the winter food of Long-eared Owls, and it is surprising to find rings from hole-roosting birds, like the tits, amongst the finches and sparrows. We just do not know how these tits are caught, but we do know that five or six years ago, our one pair of Marsh Tits in the wood was caught and eaten by the owls—their rings turning up in the pellets quite close in time to one another. Regrettably, no others have taken their place—but perhaps sensibly!

Most years, January is also the month that the Whitefronted Geese on the marsh are feeding in the fields of winter wheat or on grass nearest to the wood, and during the day disturbance by a shepherd or farmer will move the birds around. Although, not infrequently, skeins of geese will overfly the wood itself, to me the charm is in the wonderfully wild noise they make—specially mystical on a foggy day when the birds themselves cannot be seen. Wild geese, I think, are one of the birds better seen *without* the use of a birdwatcher's ever-present binoculars—rarely, save at a reserve, can you get close enough to appreciate plumage detail, but flying geese present a kaleidoscope picture set against the remote calmness of an estuary backdrop and its superb skies.

As the steady process of converting the rough grazing of the marshes into highly productive cereal and potato land continued, some concern was expressed that the geese might be driven away for lack of food—they were not expected to take to winter wheat in the way they did, and the light grazing they perform actually improves the wheat crop by increasing tillering in the way that light grazing by sheep would—so both the geese and the farmers are reasonably happy with the situation.

Towards the end of the month, in a mild, open winter, snatches of the fragile winter song of the Robin and bursts of the simple, robust, song of the Mistle Thrush are all that break the daytime silence of the wood. Mistle Thrushes sound their best in the milder days that often accompany fairly strong westerly winds—perched bravely high in the tallest trees against a sky mellowing towards sunset. In an open winter, January sees the first precocious dog's mercury shoots pushing through the leaf mould—soon to suffer if frost and snow return—and the flowering of the peculiar butchers' broom. This plant has abandoned leaves in all but vestige, the flattened stems serving both structural and photosynthetic purposes. Regrettably it is quite unusual, for besides the use indicated by the derivation of its name, as an evergreen it is admirably suited for frosting or gilding for ornamental use at Christmas, and clumps at High Halstow suffer at the hands of gipsies intent on satisfying our demands for such gaudy trivia.

In a hard winter, our evening departure from the wood is prompt and quiet—roosting birds need their shelter undisturbed—but in the mild conditions of most recent years we have had time to pause, towards the end of the month, to see the first few Herons arriving to roost in the heronry trees. Once or twice in the last ten years birds have started to display, repair nests and even lay eggs at this startlingly early date, and in one year we had the remarkable record of the first egg in January, and the last youngster leaving the colony at the beginning of October!

A delay till after dark may produce a situation that remains spine-chilling to the naturalist, no matter how many times he has encountered it. There is nothing in wildlife more human, nor more blood curdling, than the scream of a vixen—and even the strongest of stomachs must pause to take a grip on itself.

FEBRUARY

February is a month of change for the wood: a mild, open winter may be brought to an abrupt halt by a few inches of snow and sub-zero temperatures, reminding us what winters *can* be like; or alternatively there may be a spell of quiet, fine weather with surprising warmth in the sun at mid-day, and with pale rose sunsets.

Should the latter be the case, it is astonishing how quickly the mood of incipient spring grips both the woodland plants and animals. The rusty brown bracken from last summer that has since autumn impeded progress by tripping the unwary seems suddenly to have flattened. Everywhere bluebell shoots appear, pushing aside the leaf litter and making two or three inches growth in the month. Early dog's mercury may even flower—very discreetly, for the tiny green-petaled rosettes are objects of beauty only under a magnifying glass. In the more sheltered spots, soft adventurous shoots spring from elder stems—rapidly burnt off in the merest hint of a frost—and in the clearings a few flickers of gorse bloom at the start of the month have turned into a yellow blaze as March approaches.

In the evening, the vixen's scream is regularly answered by the dog fox's sharp bark, and in early morning the muddy paths show even the most inexperienced tracker the volume of animal traffic that has passed during the night. Sometimes a trail of feathers will lead to a fox kill—occasionally a pheasant but more often a woodpigeon. The soft white down feathers that insulate the pigeon in winter are scattered over an extraordinary distance from the corpse—to such an extent that one wonders how on earth they are

77

all crammed onto the live bird. Presumably these victims were roosting low enough to be snapped up on the jump by the fox, but on occasion (and indeed perhaps frequently) this may be a case of the fox tidying up in an almost kindly way behind the farmer. With their considerable acreage of brassicae, and now that cereals have displaced many of the clover-rich leys, the farmers of the Grain peninsula are jealous of the marketable quality of their cabbages and brussel sprouts and the winter day is regularly punctuated with shots as irate farmers pursue hungry pigeons. Obviously many birds are just 'winged' and limp, injured, to the wood for shelter. They cannot attain the high, safe, roosting spots and in many cases are quickly put out of their misery by the foxes.

Despite the amount of work that has been put into the study of Wood-pigeons by the Ministry of Agriculture—all of which tends to show that (as in many bird populations) winter weather, food shortage and disease combine to reduce the population naturally by the spring to the level of the previous year—farmers are naturally rather difficult to deter from forceful protection of their crop. They remain unconvinced that the level of control that their attempted slaughter achieves is only sufficient to ensure that there is enough food for all the remaining pigeons, and that their major ally, starvation, is thus hardly allowed to become effective. It is interesting to reflect, too, that the headlong pursuit of fashion and the debatable grail of increased profit margins has caused them to abandon clover pastures—the pigeons' preferred winter diet—and to bring upon themselves, to a large extent, this plague.

Fox paws, whilst basically dog-like, are more diamond shaped, and in snow a trotting fox leaves its prints in a near dead straight line, evenly spaced, showing how balanced and beautifully poised its movements are. A closer look at the paths in Northward Hill in February shows a different sort of foot pattern altogether—likened best, I suppose, to the sort of print a fairly small human baby would leave. There is a clearly marked oblong 'sole', and along its leading narrow edge are arranged five neat little toes. Sometimes a track like this can be followed for hundreds of yards, and traced back into the hawthorn thicket area of the wood. Here, in a mild February, we can find other signs of badgers—for badger tracks they are—in the form of grass and leaves, turfed out of the badgers' underground set as unwanted and elderly bed litter in the first 'spring clean' of the year. Over the last sixty or seventy years, the badgers' diggings have become fairly considerable earthworks, with many entrances, covering in one case an area perhaps 20 yards by 40. The now trampled-down mounds of soil from their excavations are often two or three feet high, and as the set entrances are normally in gullies about a foot deep, for humans the going after dark is rather treacherous.

In a mild winter, a walk through the heronry is always capable of producing a surprise. As in all dealings with the wild, this sort of visit must be accomplished infrequently and with some stealth so that the colony is not disturbed. The first few days of the month will see a few birds roosting in the trees overnight, but by mid-month the extremely noisy honkings, screechings and bill-clappering of display have reached a crescendo. Astonishingly, the broad wings and long legs of the heron do not produce quite such an ungainly bird in the trees as might be imagined. Steadying themselves with wings outstretched (like a tightrope walker and his pole), the herons pace sedately about on their spidery toes in the topmost twigs of the elms, raising their crests and throwing back their heads in ecstasy on meeting their partner. Incidentally, this sexual arousal also brings a rosy flush to the normally yellow bill. The pair bond is sealed by copulation—again carried out in the treetops, and nesting proper starts.

Most often, the majority of the nests in the herony will have survived the winter's gales, and most birds, if they too have survived the hazards of winter, will return to the nest they used last year. Naturally, some changes of partner occur, and new birds enter the breeding phase, but new birds, whether newcomers from elsewhere or Northward Hill youngtsers breeding for the first time, will usually start much later in the season. For many years an old, very pale—almost white—female occupied what was both the largest nest in bulk and, so far as we can see, the nest best sheltered from the elements, especially cold winds. She was always the first bird to be sitting tight, and came to be regarded as a sign that spring was on the way in earnest, although on several occasions she could later be seen sitting with a drift of snow on her back!

Nestbuilding thus becomes a repair job rather than a creative one, but there are complications as the presentation of sticks to the female is part of the ritual of courtship. Males of calibre clamber around on high, seizing often quite sizeable branches (an inch or more in diameter) in their bills and attempting, sometimes with success, to snap them off. High in the trees, if you are a heavy bird, this operation can be tricky and is rarely accomplished with any grace. The chickens amongst herons descend to the clearings and root around for dead timber, returning just as proudly as the others bearing gifts to their female. Inevitably for many birds the temptations of a nearby untenanted, or briefly unguarded, nest are too strong to withstand, and pilfering is rife.

The number of occupied nests increases during the month, with an accompanying increase in activity and cacaphony. Towards the month's end, the most impressive sight of all is to see the females tucked down as low in the nest as possible riding out a premature March gale, all head to wind. Some of the nests in more slender trees have been measured as moving through an arc of more than 15 feet in a good gale—presumably a terrifying experience for those birds sitting.

Mild weather also provokes a fairly substantial amount of song: the Mistle Thrushes—perhaps ten or a dozen pairs scattered through the wood—are now hard at it, and some will have built nests—bulky structures of moss and leaves (and all too often, today, streamers of torn polythene bags) set in the angles of branches, or often atop a thick branch at an elbow. Song Thrushes, although less numerous than Blackbirds, at this time of year outdo them in the dusk chorus, and only towards the end of the month does the rich fluid song of the Blackbird oust the more stilted, repetitive phraseology of the Song Thrush. Taking time off from foraging with other tits, Treecreepers and Goldcrests, male Great Tits begin stridently to stake their territorial claims in the wood usually centred on one of our nestboxes. The tee-chah tee-chah tee-chah is often the precursor of the other disyllabic singer, the Chiffchaff.

How different this story is if, as not irregularly happens in eastern England, February turns really savage. For a start, given six inches or a foot of snow, the journey out to the wood by car takes on a new and interesting dimension: what is more dangerous, the other traffic on more major roads, or pressing on regardless on the untreated minor ones? To my mind the last part of the journey, including the hill, along narrow, twisting, high banked lanes in deep snow is one of the welcome experiences of a winter, keeping the basic motoring instinct alive and alert.

To the birds this picturesque covering must present a very different, and totally unappealing aspect. The ground on which so many of them feed has vanished, and should there have been a glazing frost, or a thaw followed by a snap freeze, the buds and twigs, in an impenetrable but glass-clear ice casing, are inaccessible to the flocks of tits and Goldcrests.

The first signs of impending trouble appear before we reach the wood— Lapwings and Skylarks, many probably coming from the Continent, are passing westwards overhead moving towards warmer areas on our oceanic seaboard. Woodpigeons, too, are on the move. Relatively few of these will be Continental birds, to judge from the ringing recoveries, but mostly local birds looking for fields of kale or brussels where the strength and height of the plants has broken the snow cover and exposed some eatable greenery. Over such fields the fusilade of shots increases in tempo, and flocks of increasingly hungry birds wheel over the sky.

In the wood, the thicker canopy areas amongst the oaks, and the very dense hawthorns, keep some ground free of snow, although some hawthorns may collapse completely under the weight of the accumulated snow. In these snow-free areas are crowded all the thrushes that have previously used the wood only as an overnight shelter, leaving each morning to forage over the fields, sometimes up to a few miles away. They mostly came in from the Continent during the autumn as part of their regular migration pattern, and now join in the search, and competition, for food with the birds that spend their whole life as Northward Hill residents. Sometimes hundreds of Blackbirds, Fieldfares, Redwings and Song Thrushes will work in crescentic waves through the leaf litter, turning over everything time after time, and making, just by their rustling, a noise like a herd of foraging pigs. When a small invertebrate is found, a tremendous battle may develop if the unfortunate discoverer does not devour its find immediately—the energy that is wasted in squabbles of this nature, with birds leaping high in the air and feathers regularly flying, and when one bird encroaches on the much-reduced-by-circumstances hunting area of another, only contributes to the general problems.

In these conditions, overnight losses of weight—caused by the metabolism of stored fat to maintain body warmth and functions—may approach 10% of total body weight, and a bird such as a Blackbird can generally only manage a week of really lean days before reaching the point of no return. It is for survival in just these conditions that so many of our birds increase in weight during the winter: a striking example of the role that man can play is offered by comparing 'town' Blackbirds—helped over hardship by benevolent housewives with ample scraps—with their less fortunate country cousins. A good midwinter weight for a healthy Blackbird is about 130 grams, compared with a mid-summer weight of about 100 grams. Luxury-fed towny birds in north London maintained a 140 gram level right through the 1968 cold snap, while our birds in Northward Hill—a National Nature Reserve, mark you—rapidly lost weight—falling to an average of about 100 grams after a week and to 80 grams or below after two weeks of snow cover. At around the 80 gram mark many of the birds were pathetically feeble and it seems that this is about the point-of-no-return for many Blackbirds—corpses were increasingly obvious. Redwings had suffered even more severely after a week, when their normal 60–70 grams had fallen to about 40 and many could be seen pathetically, but literally, dying on their feet incapable even of flight. Even though the thinking naturalist must accept this population levelling as an essential for healthy populations of a variety of species—for 'safety's sake' most species over-produce young during the breeding season, and the oldest and least fit, physically or otherwise, perish during the period before the next breeding season—it remains hard to remember the general principles of evolution and survival involved when confronted with the actual mortality.

Even when snow does not accompany lengthy severe frosts, there are problems for the thrushes in breaking through the frozen crust of soil to get at the small animals for food. Partly the problem is a downwards migration of these small animals to slightly warmer soil, but at least as important is the mechanical difficulty. In cases where severe frost extends over two or three weeks, it is the thrushes with the shortest beaks that suffer first—Redwings, then Song Thrushes, Fieldfares and lastly Blackbirds and Mistle Thrushes.

In these conditions, while the thrushes suffer, the tits tend to seek their own salvation along the lines of the urban Blackbirds. We were lucky enough to be netting on the fringes of the wood bordering the village one day at the onset of a snowy period. The next chapter gives some idea of the territorial activities of tits during the winter, but on this occasion, shortly after the snowfall started, we began to catch birds, all moving towards the village (and presumably towards well stocked bird tables) from all parts of the wood—clearly getting out the quickest way. This departure of the tits to the village in such conditions is a regular and fascinating feature, but although the feeding may be better and survival apparently more assured, other, previously unfaced, hazards occur, and we have an increase in the number of recoveries of our marked birds, caught by domestic cats making the best of unexpectedly easy hunting.

MARCH

For some years past, a small research team has been studying the biology and the population dynamics of various woodland birds, especially the tits, in Northward Hill. Knowledge of this sort is essential for proper conservation planning. During the summer months we shall see more of the territorial activities of the breeding season, but studies continue through the year, and surprisingly enough some aspects of territoriality do persist throughout the winter. For these studies, naturally the birds must all be recognisable to the researchers as individuals, and this is achieved by placing a small metal ring with a serial number—just like a car registration number—on one leg. These rings are so light—weighing the relative equivalent of somewhere between your signet ring and your watch—that they do not interfere with birds' normal lives at all. Yet, because of the individually recognisable number, the birds' histories and movements can be documented even should they leave the wood and later fall victim to hazards like cats or road traffic.

The most common technique for catching the birds is to use a fine net—roughly of the texture of a hair net, about ten feet high and from twenty to sixty feet long. These nets are called mistnets because of their near-invisibility when set against a background of trees or bushes, and flying birds, failing to see them, fly into them. The fineness and elasticity of the nets allows the birds to slide gently down into pockets behind the supporting strings, whence they can be quickly and easily removed by the ringers. After removal from the net, the ring number is noted, and on each occasion the bird is weighed and state of plumage is recorded. In this way, a lot can be found out about the length of life of the various individuals and their movements around the wood, and on occasions outside, can be traced. Plumage details can tell the experienced eye much about the age and the sex of the birds—vital information in studies of the workings of bird populations. Usually the catch during a day exceeds fifty individuals, and on a busy day may near 200—enough, with the detailed paperwork, to keep a team of two or three pretty busy.

Handling birds is no simple matter. Some of them—especially the tits—are well equipped in an anti-personnel sense, and an attack by a practised Blue Tit at the quick of a finger nail in cold weather has to be experienced to be believed! Blood is often drawn. On the other hand, birds are small, fragile creatures compared with the human hand, and an orderly and smooth arrangement of the feathers in winter is absolutely vital to maintain decent insulation. Thus all bird ringers have to undergo a lengthy period of training by experts, and must hold both a Government Licence and a Ringing Permit issued by the British Trust for Ornithology (who control ringing in Britain) before they can embark on studies of their own.

Ringing was 'invented' at the turn of the century primarily so that the migratory movements of birds could be studied: this is how we know that our Swallows winter in South Africa, and that our young Manx Shearwaters spend their first winters at sea off the coast of Brazil. Whilst the dossier of information on migration is still being drawn up for most species, increasingly ringing is being used as one of the most valuable research tools to provide the accurate biological information that is conservation's strongest weapon. Reports of ringed birds tell us not only where they go, but how long they live, and when and why they die. So we may measure the impact of the wildfowler on ducks and geese, of the Belgian cage-bird fanatics on our migrant finches, and of the Mediterranean delicatessen on our warblers. This last reveals a quite shocking state of affairs: to satisfy the appetite for luxury tit-bits (*not* basic food needs) in countries like Italy and Malta, huge numbers of the Blackcaps and Willow Warblers reared in Britain each year, including some from Northward Hill, perish on bird-limed sticks or in snares set by local bird catchers.

But to return to our studies of tit territories: following the fiercely territorial mid-summer period when an area surrounding the nest site is defended, territorial behaviour as such seems to break down—perhaps influenced by the collective stresses of shepherding a large family about and keeping them well fed. In late summer these family parties coalesce but as winter sets in they become smaller again, and much less mobile. Our ringing sites have been in much the same places, scattered throughout the wood, for the last ten years, so it has been possible to see how many birds are moving about at any time. Below is a stylised diagram of the situation in mid-winter.

It seems that when the large parties break up at the onset of winter, both Blue and Great Tits space themselves out fairly uniformly over the whole wood, and most will then move only within a fairly closely defined area,

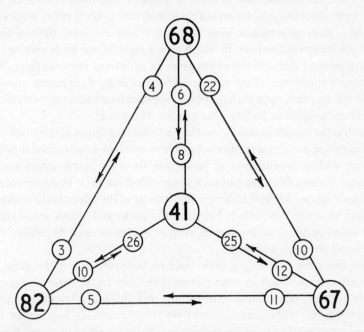

INTER-SITE MOVEMENTS—the large circles represent ringing sites, the three outer ones about 400 yards apart, the inner one about 200 yards from each of the others. The figure within the big circle is the percentage of birds caught at that site during a winter that were caught nowhere else. The figures within the smaller circles indicate the percentage of birds subsequently caught at the sites indicated by the arrows. Clearly for the peripheral sites the faithfulness to one site through the winter is very strong, and only at the central site, perhaps on the margins of the other three 'territories' did any substantial overlap occur. It is interesting in that in every case, the major 'exchange' route follows the line of 'least resistance'—there are uninterrupted communication pathways through the vegetation, and the two sites in each case are at about the same contour level, with no rising ground between them.

perhaps less than 400 yards across, whatever the temptations in neighbouring areas. This group-wintering-area does not seem to be actively defended as is the breeding territory, but the boundaries in most cases seem reasonably sacrosanct through the winter. In the following winter, each bird may choose a different area, or may remain as before. About the only thing that will move the tits out is a period of snowfall—as related in the February chapter—but once the hard period is over, the birds return to their favoured areas once again. Apart from ensuring reasonably even distribution of birds and their food supply, it is difficult to find a reason for this loosely territorial behaviour.

One of the fascinating aspects of ringing is the sheer pleasure of such close contact with the bird and observation of the details of feather structure and colouration. An examination of various species in the hand reveals far more of the marvels of the adaptations of their structure to their many ways of life than binoculars or armchair work with a textbook ever can. Beyond that, a long-term study such as ours, in which we see many of our birds once or twice a month or more through their lives gives an additional intimate facet. Newcomers and youngsters of the year are recognisable by their recent numbers, but old friends, with rings perhaps put on between four and seven years before, always raise a welcome feeling—'he's going strong still'.

March is the month to dawdle on the southwestern slopes of the wood: each year it comes as a surprise to discover just how much more advanced the plants are here, with a combination of protection from the northeasters and the additional warmth from the sun on a slope with this aspect. Bud burst may be up to a fortnight earlier and under the thin canopy of the ash trees the scattering of violets shows to best effect. Fringing the paths and grassy areas are the bolder splashes of colour from some very robust primroses, benefiting from the rich soil and deep leaf litter.

Apart from the bluebells a little later in the year, this is the time that botanically the wood is at its most picturesque. Over in the main clearing, the gorse is well and truly ablaze with colour, and strikingly bright against the leafless darkness of the surrounding trees, the blackthorn has turned from sombre leaden patches to drifts of unbelievably solid white blossom. Simple, common plants these, functional so far as the birds are concerned winter through, and now both emotionally and aesthetically satisfying as typical of an English spring. Looking out over the marshes, by now there is nothing to be seen of the geese, but the winter-visiting Short-eared Owls may still be in evidence bouncing across the remaining rough pasture. All traces of ploughed land have vanished as the winter wheat accelerates its growth, and the marsh is all-green again as it was before drainage became so popular. If we are lucky we might see either a Marsh Harrier or a Hen Harrier quartering the marshes and the reed beds of the fleets on characteristically stiff wings, occasionally dropping in an attempt to capture an unwary water vole or Moorhen. When they drop, you can see just how long their legs are, and realise how important this extremely long reach is to a predator, much as you wouldn't realise it from the usual pictures of birds of prey in action.

This activity on the marsh has its counterpart within the wood: the small mammals, voles, fieldmice and shrews, are moving about hungrily in the unaccustomed warmth of spring, and naturally enough the predators are taking advantage of this. Towards the evening the Barn Owl from down in the heronry will be about—a pale ghostly figure, but surprisingly difficult to distinguish in really poor light from one of the Long-eared Owls putting in a late appearance before returning to the Continent to breed. Our Barn Owls have nested fairly regularly in a hollow, broken-off side branch of one of the heronry elms—an uncomfortable climb of some twenty feet up a slender lichen-covered trunk with no vestige of foot or handholds and a very poor view down into the dark of the nest when you get up there. Far better for access when they nest in one of the nearby barns on regularly-stacked hay bales, but they are very mysterious over their comings and goings and if it is not in the heronry, the nest is always a problem to locate.

In the vicinity of the wood in most years are two or three pairs of Kestrels, nesting sometimes in the wood, sometimes in old isolated trees or farm buildings. Generally at least one pair will have territorial aspirations over parts of the wood, and the main clearing is a favoured place for the wonderful courtship display. We tend to regard the Kestrel as a master of the air when it comes to hovering on flickering wings—a real *tour de force* and worth every minute spent in close observation through binoculars. Notice how the head is rock steady all the time, unaffected by the adjustments of body and tail, and wing beat, that keep the bird in position even in a blustery wind. Early on a March morning, looking cautiously out over the clearing from part-conceal-ment under overhanging bushes, we can realise that the Kestrel is just as much a falcon as is the Peregrine. Overhead a pair are displaying—and immediately the striking difference between the tiny male and the robust female is apparent. In the Peregrine, this size differential is such that the male must court with caution or else be mistaken for a delectable prey item! But back to our Kes-trels: a full courtship aerobatic display may go on for over an hour if uninter-ruped—with birds alternately climbing high, kek-kek-kekking shrilly before plunging down in a dramatic stoop at the other, which of course dodges nimbly at the last moment. Sometimes mock combats result, with the birds tumbling over and over in the air, occasionally with their feet interlocked. The continuous calling may summon a neighbouring pair, or perhaps an inquisi-tive non-breeding youngster from last summer—and then the combat will intensify and become more real than apparent, as both birds combine in attacks to repel the intruders.

By now, of course, most of the migrant thrushes, the Fieldfares and Red-wings, will have moved off eastward, and no longer do huge Starling flocks commute from the roost in the wood out to the marsh. March, at its close, is to some extent a month in limbo, for most of winter birds have gone, and the bulk of summer birds have yet to come. The herald of the summer birds is here, though, and on every warm March day we keep our ears cocked for the one summer song (beside the Cuckoo) that needs no re-learning each year. Sure enough, there it is, a Chiffchaff, lisping away in one of the willows—and if memory serves correctly, is that not the exact tree that last year's first Chiffchaff sang from?

APRIL

The wonder of coming summer influenced primeval and medieval men strongly. Except in remote areas it is left, in our modern—sometimes over-civilised – society, to those with an interest in nature to appreciate such things. In Northward Hill, with all our marked birds, we have the benefit of an additional dimension. Not only was that first Chiffchaff singing in the same tree as last year's first bird, but when we caught it a few hours later we were able to confirm, from its serial number because it was already ringed, that it was indeed *the same individual* that had held the territory centred on that huge old willow throughout last summer. Since we last saw, handled or heard it, this very fragile and rather drab seven grams of bird (about one quarter of an ounce) has wintered south, and probably well south, of the Mediterranean. And in a head smaller than the size of my little finger's last joint is the control centre not only of the everyday functions of the bird, but of a navigation system capable of getting the bird back to its territory with this precision!

All through the summer we will be rediscovering this aspect of bird migration: not only can the young Cuckoo reach Africa unaided by its parents (except that in the genetic sense they have contributed to its instinctive abilities) but the whole business involves meticulous accuracy. From the marked birds we recapture each year in the wood after they have spent the winter in Africa we may build up a picture in accord with that emerging nationally. Adult birds—that is those that are breeding—if they survive the rigours of the migration journeys (even this seems close to a miracle!) are most likely to return not just to the same county, but to the same wood, and the same area in that wood even when the total area is only 130 acres. By the same token, allowing for the inevitable regularity of deaths and for occasional 'divorces', the Swallows in your porch or garage are likely to be the self-same birds from one year to the next. The young birds reared in the wood, however, seem to distribute themselves more widely. Because the mortality amongst the young is so much higher, and because (obviously) it is easier for us to study a bird that returns precisely to an area where ringers are working regularly, it is more difficult to draw sound conclusions on the evidence available. Nevertheless, it would seem evolutionarily sound to suggest that the circumstantial evidence does point to a generally wider distribution of returning young birds about to embark on their first breeding season, for in this way the dangers of inbreeding (pooh-poohed, perhaps with justice, by some biologists today) will be avoided, and any local deficiencies such as a breeding failure for one reason or another will be remedied, rather than being followed by a local extinction of that particular species.

Not only will late March have brought us our first Chiffchaff of the year, but it will have allowed us to listen with some ease to two of the songsters in the wood whose song is so fragile as to be drowned when the full choir of summer gets going at full blast. Early April is the same, and an afternoon wander through the oaks should yield Treecreepers, whose song is a descending spindly trill and final flourish, and Goldcrests, whose song follows the same pattern, but is scratchier and even higher pitched. It is said to be a sure sign of increasing age when the birdwatcher's eardrum loses its sensitivity to these high frequency notes.

The Treecreeper population of High Halstow probably amounts to twenty or thirty pairs, but they are inconspicuous birds and numbers are difficult to assess. The numbers seem, on subjective assessment, to have been relatively steady for some years now since the last really severe winter of 1962–63. There is a relatively restricted song period early in the spring, for the Tree-creeper is an early breeder. Probably most pairs only raise a single brood, but a few will attempt a second brood and a further brief burst of song activity will precede this at the beginning of June. In Northward Hill, this population stability persists despite what has always been an abundance (and what now, because of Dutch Elm disease, has become a superabundance) of suitable nest sites behind flaps of peeling bark. Possibly food supply is the limiting factor beyond this. An April walk through some of the oak areas may find us a nest or two: although the undergrowth is thin here, it is still difficult to move quietly because of the deep leaf litter and the twigs hidden under the leaves, ready to crack sharply under a clumsy foot. The reason for this caution is twofold: the commonsense one that if you move quietly and unobtrusively, you see much more anyway; and secondly, although it would (just) be possible to move from tree to tree looking for favourite nest sites, it is much easier to find a pair of Treecreepers in the process of nestbuilding, and let them lead you to the nest itself.

At Northward Hill, the commonest nest-site is behind a peeling flap of bark, perhaps six inches or so across and an inch or two clear of the trunk. In this narrow slit, the boat-shaped nest of fine twigs is lodged and the eggs laid. Most often, not only is there a 'front door' slit entrance, but an emergency exit above and behind the nest too—a sensible precaution for the sitting female to avoid marauding predators like squirrels and weasels. The very nature of the nest site under rotten bark, and the tenuous nature of the nest supports, mean that we must be most careful when inspecting the nest. Perhaps the best technique is to use a small torch and either a dentist's mirror or a fragment of hand mirror mounted at an angle on a short holder. Thus the bark flap need not be touched at all.

Goldcrests show a very different picture—all over Britain since the severe setback of 1962–63, when snow and ice dropped the population by more than 50%, their numbers have been climbing steadily. Although conifers are the preferred habitat, Goldcrests are now expanding into deciduous woodland —perhaps due to sheer population pressure. From being regular winter visitors in the early sixties, late in that decade we had males singing through the summer months in the hope of attracting a mate, and following the mild winters of the early seventies a few pairs are now breeding each year.

As the month develops, so more and more migrants appear. For the earlier ones, it is almost possible to predict the sort of day (or rather night, for most birds are night migrants, resting and feeding by day) on which arrivals will occur. Ideal are calm, clear nights with little wind, or for us at Northward Hill, light breezes with a touch of north or east in them. Little movement will take place in heavy cloud, or in blustery wet weather. Sometimes April will catch a hangover from March in the way of a few squally days and nights at a stretch, and somewhere to the south of the track of the cyclonic weather (or depression) across the Atlantic, the hordes of migrants will be held up. While the dam holds, the wood is quiet—song from the thrushes, Blackbirds and other residents, but even the early arrival Chiffchaffs have fallen silent— perhaps because they need in these unfavourable times to spend much of their day feeding sufficiently to maintain life. Ultimately, the dam must break, and the metaphor is not misplaced, for at the first sign of a suitable spell of weather, the interrupted migrants press on northwards even more anxiously.

Such a situation causes an almost annual 'event' at the network of coastal

Bird Observatories—especially those along the southern approaches. Sometimes the bushes, and even the grass, seem to be a moving carpet of newly arrived but tired and hungry birds, seeking to feed up as quickly as possible so that further progress is not impeded. The term 'fall' that is applied to such an arrival is not at all inappropriate. In the wood things are little different—after a few stormy sullen days, a calm mild and fine morning arrives, and everywhere there are Willow Warblers singing! The other species that tend to arrive conspicuously like this are the Turtle Dove, the Whitethroat and the Lesser Whitethroat, but although their advance guard arrives at Northward Hill in April, the bulk of these three species usually will not reach us until early May.

Other, less obviously dramatic arrivals during April will be the Blackcaps (again, the older residents will come straight in to the bushes of last year's territories), followed after two or three weeks by the Garden Warbler. For some reason, the Northward Hill Cuckoos never achieve a calibre meriting a mention in the correspondence columns of *The Times*.

While the Mistle Thrushes are now feeding young, Blackbirds and Song Thrushes are well away on the first of what, for some pairs at any rate, may be four or five broods if the season remains favourable right through, and the remainder of the residents are at least building nests, some other species appear as if from nowhere to become strikingly apparent. A number of these are with us winter through, but are then much less significant for some reason: the best examples are Linnets, Chaffinches and Yellowhammers. Perhaps the difference is that they roost, relatively quietly and certainly inconspicuously, in large numbers of which only a few pairs remain in the wood to breed. Certainly the males of each are conspicuous enough in April—the Chaffinch largely by its superb song from within the canopy, the other two more because of the prominent bush-top song posts that they occupy in the clearings. While the Linnet has a pleasant song, it is just a little jingly and unformed, while surely only a female Yellowhammer could find the dry, rasping notes (which to my mind bear little, if any, resemblance to 'a-little-bit-of-bread-and-no-cheese') of the male at all attractive.

To close the month, we should walk down into the heronry to see the state of play there. This will involve us in cutting across the big clearings in amongst the oaks, which are now at their most chocolate-boxy. Bluebells, in dappled sunlight under trees, have a peculiarly English beauty. Those who know small west-coast islands, treeless but bluebell covered, will appreciate the impact that this superb plant can have *en masse* in the open. In Northward Hill we are doubly fortunate, in that we not only have bluebells under the trees, but we have them in the open as well. Ringed by oaks, the longest stretch is over 200 yards of unbroken bluebell—usually well synchronised, this solid sheet of blue with the yellow-green of spring leaves as a backdrop is one of the most beautiful sights I have ever seen, and even though next spring will be the 25th anniversary of my first experience I look forward to it as much as I did the second.

Following an open winter, even in late March a quiet walk beneath the nests might allow us to hear the tick tick tick . . . of newly hatched Heron chicks. Certainly this is the heronry sound of April, and as the chicks grow their voices deepen to a tock tock tock—produced with metronome regularity when hungry. Searching under the nests from which noises emanate will give us a chance of finding the hatched eggshells and assessing how many chicks have appeared. Heron eggs are only a little bigger, if anything, than those of a chicken, and are bright pale blue. Those that have successfully hatched can be identified by the even chipping round the largest 'equator', and by the presence of a transparent and now drying membrane, containing many brownish blood vessels, on the inside of the shell.

So, by gentle degrees, the pattern that describes summer for us in Northward Hill has been established—the Chiffchaff, the blackthorn blossom, the bluebells, leaf emergence, Willow Warblers in quantity, and now a rapidly increasing number of nests with young in the heronry.

MAY

Hand in hand with the ringing work throughout the year goes the study of the nestbox colony of Blue and Great Tits. Between fifty and seventy boxes have been in position now for ten years, and regular rounds to inspect the contents and their progress start late in March or early in April. The great majority of the boxes are in or near the oaks on the north-facing slope, but a sub-unit is positioned over on the southerly aspect.

We have used a very simple design for ease both of operation and mass production, and the boxes have proved astonishingly long-lasting in the field, especially considering their inexpert construction and the material used—floor-boarding from houses in the process of demolition! The boards are 6 in by $\frac{1}{2}$ in or $\frac{3}{4}$ in, and the basic dimensions are 8 in high (rear) 7 in (front) by 7 in to $7\frac{1}{2}$ in wide depending on the thickness of the wood. The hole is positioned in one side, and the lid, which *must* be removable, overhangs each side by an inch or so. The lid may be secured by a block and a hook and eye. An entrance hole diameter of $1\frac{1}{8}$ in is adequate for Blue Tits, and in our case for Great Tits also. It will shortly become apparent that a protective metal plate surrounding the hole is advisable in the conditions at Northward Hill!

The boxes are fixed on a variety of trees, oaks naturally predominating, at heights between four and twelve feet. Even though access to the wood has

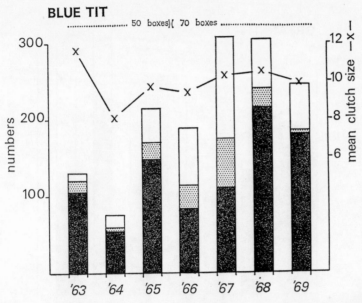

BLUE TIT—the seven-year record of productivity from the nestbox colony. There were 50 boxes until 1965 and 70 thereafter. For each year, the height of the column measured against the left-hand axis indicates the number of eggs laid. Egg losses are indicated by the white areas. Usually these are due to predators like the Tree Sparrow, Great Spotted Woodpecker, Squirrel or other small mammals, and they may be (as in 1967) quite considerable. The average number of eggs laid has been about 209, and the *average* egg loss each year is nearly *one third*. The spotted area indicates the number of young birds lost after hatching, and the darkest area the number finally fledging. Losses of young average about 15% of the number hatched, and on average about 104 young are reared successfully from the boxes each year—less than half of the eggs laid. Generally the same predators are responsible for the losses of young as take the eggs. The mean clutch size for the year is shown above the histograms, and works out at about ten eggs.

always been restricted, human disturbance is such that except in the most concealed positions all the boxes need a ladder for inspection—a nuisance as ladders are not only heavy but impede progress through thick undergrowth! The cardinal rule that we have followed in positioning the boxes on tree trunks is that they are not exposed to extremes of sun or wind, and are not placed in the path of streams of water coursing down the trunk in heavy rain.

Nestbuilding will sometimes start in late March with the collection of a few pieces of moss, but during April building gets under way in earnest. Occasionally Great Tits will occupy really large boxes designed for Little Owls or Stock Doves, and on one occasion a pair collected nearly one cubic foot of moss to fill the box to an acceptable level, and to bring the nest cup itself—tiny and lost in an expanse of moss—near to the hole. The first

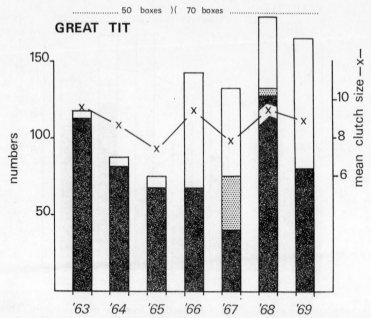

.... 50 boxes   )(70 boxes ....

GREAT TIT

GREAT TIT—the layout of the diagram is the same as for the Blue Tit. The average clutch appears as between 8 and 9 eggs, smaller than that of the Blue Tit. The losses of eggs are variable, averaging 38% of the 129 eggs per annum, and strangely enough largely due to nestbox 'takeovers' by Tree Sparrows. Great Tits seem to be better able to defend their young, however, and losses of young are much lower than for Blue Tits at less than 1%. Only in 1967 (the 'weasel year') were losses appreciable.

foundations may be wholly moss, or there may be a framework of dead grass or rootlets. Certainly in the wood there is abundant moss, very loose, for collection, and the main bulk of the nest is moss. Usually in a corner of the box a 'cup' is formed crudely in the moss, perhaps $1\frac{1}{2}$ in across. At this time the eggs are laid—usually before the process of lining the nest is properly under way.

Great and Blue Tits lay an egg a day, and at Northward Hill almost invariably the female will conceal the eggs in the loose nest fabric until the clutch is nearly complete, when all the eggs will be rooted out and gathered together in the cup. During laying both birds, but especially the male, will have been busy gathering the nest lining—soft and insulating. In the case of the Blue Tit, the lining is almost always of feathers, and the regularity of this allows a preliminary identification of the nest even before the adults are seen. The feathers are usually obtained from corpses, and thus tend to be from the leftovers of fox kills—usually Woodpigeon or Pheasant—the former with lots of snow white down, the latter less practical (because of poorer insulation properties) but far more attractive because of their rich colouration and metallic sheen. The Great Tit in contrast generally lines its nest cup with animal hairs: sometimes these will be sheeps' wool or cattle hairs collected off barbed wire (very rarely horsehair nowadays), but more often tufts of fur from dead rabbits or from fighting hares are used. Occasionally a source of fox fur is found and used. In the latter instance, as you stand at the top of the ladder peering in, it is abundantly obvious that this is the case because the fur retains the characteristic, and not pleasant, persistent odour of foxes. Perhaps it is just as well that birds have little or no sense of smell, for the youngsters keep the taint for a while after they have left the nest.

So through April, May and early June, we patrol our nestbox round, noting the clutch size or brood size and its state of development. Fortunately for our study, the tits are very amenable, and are not easily disturbed. So, with practice and due caution, the sitting female can be removed from the eggs by hand, her ring number is read and she is weighed and then gently replaced. If the feathers over her brood patch—the bare skin of her belly, richly supplied with blood vessels and the prime means of supplying the warmth to incubate the eggs—are blown gently aside, as soon as she feels the eggs in contact with her body she will promptly settle back down to work. When there are young-sters, we use automatic trapdoors on the holes, which close after a feeding parent has entered. He or she can then be taken from the box and examined and identified. These processes obviously only take a few minutes, so that the daily routine of the family is not disrupted. Great Tits and Blue Tits are amongst a very small number of birds for which this sort of intimate study (allowing a year-round investigation of their biology) can be carried out, and thus its value is very great. Most birds are not amenable to study during the breeding season, other than seabirds, but these are difficult to study during the rest of the year.

We also need to investigate the adverse factors operating against bird populations if we wish to understand fully their dynamics and to be able, if necessary, to take appropriate conservation measures. Often on our rounds of our nestboxes, we will find the odd box that has been attacked by a grey squirrel or by a Great Spotted Woodpecker. The latter may come as a surprise, but over the years they have been just about the steadiest adverse pressure. Somehow they recognise the box for what it is—perhaps observing an adult entering—and then they will cut their way in, by enlarging the hole if possible, or by excavating (a job to which they are ideally suited) a new one in the other side if not. The eggs or young are then eaten. As the tit nestlings grow larger, so they get more vociferous for food and tend to jump up to the entrance when they sense a parent coming by the shadow across the hole. Again, the Great Spotted Woodpeckers take easy advantage and, landing on the side of the box, they seize the hopeful but hapless chick as it jumps up to meet them.

Another surprise as a predator is the Tree Sparrow. Despite the courage of both Blue and Great Tits in defence of their nests (often a parent will perish in the box rather than flee) outwardly rather innocuous birds like the Tree Sparrow will take over the box at any stage and build in it their own untidy nests. In North Kent in general, Tree Sparrows are on the increase, and at Northward Hill in particular they are spreading widely since the clearance of many of the old pollarded willows from the marshes that accompanied the change from grazing to cereals. Late in the summer an investigation of the foundations of the Tree Sparrow nests shows the toll of tit eggs lost, and on occasion even the battered and now-mummified corpse of the original owner, still over her addled eggs.

The young tits, and indeed the success of the whole tit population, are closely dependent on the size of the winter moth caterpillar crop on the oaks and the elms, and on the timing of the caterpillar hatch. Usually, and most mysteriously, both the timing of laying by the tits, and the size of the clutch, are closely correlated to the caterpillar crop and its timing—how the tits know in advance what to expect, and when, is difficult to imagine. It may be that common factors of winter temperatures, spring daylength, spring temperatures and perhaps other things govern both tits and winter moths, but the value of the close tie-up is illustrated abundantly clearly when the system fails. Walking through the wood when there are young in the boxes you can hear the anxious, shrill piping for food, and on our nestbox round perhaps we will find three or four families cold and dead—simply from starvation.

Another predator may hear, probably better than we can, the cries of the young tits, and as a very capable tree-climber he is quickly up to the box. He can easily manage to climb through the hole, and a very messy butchery ensues. It seems very probable that after a few successful raids, the weasel comes to associate nestboxes with food, and in one caterpillar-failure year, we had 26 boxes in succession raided over one weekend. The butchery was wholesale—far more than the weasel would need for food—and we lost over 200 young tits and eggs. In one way, this story has a happy ending. In one of the boxes, of Great Tits, on our inspection round we detected movement under a pile of headless corpses, and extracted from the carnage one apparently uninjured, though much blood-stained, youngster. In the circumstances all that we could do was to find him (as it turned out) some foster parents. A quick look through the field notebook records showed us a distant box in the wood which should have youngsters of about the same age. With some trepidation we transferred him, after a clean-up, to a nest with ten new brothers and sisters, hoping that the foster parents would be so busy that a 10% increase in the family appetite would go unnoticed. So it turned out, and Ring No. HH92022 fledged successfully, to be encountered by us regularly over the next couple of years—each time with something of a glow of satisfaction, not only for being benefactors but for the privilege of participating in a study as intimate as this.

Appropriately enough, the whole of our diary for May has been devoted to the nestboxes—because it is our overwhelming concern to keep pace with the breeding season. Carrying a ladder round the boxes, climbing up to inspect each one, weighing, measuring, carefully counting eggs and the recording all take time, and ringing perhaps 300 youngsters on a good day (they seem almost as active and leggy as large pink spiders!) leaves the fingertips as raw as the leg muscles are sore. Perhaps next month will allow us the opportunity to reflect on other aspects of summer in a more leisurely way.

JUNE

May saw the arrival of the latecomers of summer so far as the migrants are concerned—the bulk of the Whitethroats and the Turtle Doves, and the Cuckoo. The descriptive name 'harbinger of spring' could hardly be less accurately applied than to the Cuckoo—if ever a bird waited until the last chance of a frost is well out of the way, and until all the preparations (in the shape of nests of Dunnocks and Warblers waiting to receive its eggs) for its arrival are complete, it is the Cuckoo—laggard of spring.

From May until August, the wood will provide the joy of a continuous background murmur of Turtle Dove song. Absolutely characteristic of warm afternoons and evenings, the steady, monotonous purring continues, confusing the observer trying to locate the singing male. The sound is far carrying and slightly ventriloquial, and in an area like the clearing, there are so many pairs located round its edges that the air seems to be *made* of purring. Every so often, a male will climb up on a display flight, clapping his wings under and above his body as he goes, to descend on parachuting wings of bronze displaying his black-and-white tail pattern. The Woodpigeon has a slightly more varied, but no less monotonous song—coo . . . *COO* . . . coo—and uses a similar display flight to mark out its territory. This is well worth watching, especially in an expanse of conifer plantation, for here the Woodpigeon population is likely to be high. I doubt that you would ever describe the Woodpigeon as a conspicuous bird, but against the sombre background of the pines, the white collar and broad white wingbars are strikingly visible—even at ranges of half a mile or more—as the bird parachutes down. The Stock Dove—on the increase again now in Britain—is a much less demonstrative bird but its call—a deep and very penetrating oo-hoo, with the emphasis on the second syllable—has a little of the foghorn quality of the Bittern's boom about it.

In common with several regularly-watched woodland areas in southern England, rarely do we miss a May or June without a Golden Oriole. In some areas in Britain these fabulous birds actually breed on occasions, but so far as we are aware, the nearest we have come to this was last year, when for at least three weeks we had two males singing in territory against one another—alas, it appears that there were no females about to hear them. Still, we look forward to the delightful shock one year of finding one of their rather Starling-like youngsters in a mistnet by the pond! As songsters, of course, they are as glorious as their rich golden-yellow and black plumage. Despite this apparently glaringly obvious colouration, put a male Golden Oriole in the dappled shade of the canopy of an oak or sycamore, with the yellow tinge of spring still in its leaves, and it vanishes. The melodious, fluty whistling, with its characteristic chwee-loo-whee-oo ending continues, but sometimes an hour of eyestrain with the binoculars is necessary to obtain even the most fleeting glimpses. Interestingly, Orioles are reasonably common quite far north on the Continent, and one of the winter-visiting Starlings to my garden in Hertfordshire in the last two years has, arch-mimic that it is, reproduced snatches of Oriole song. This can be regarded as verbal confirmation of the ringing recoveries that indicate that many of our winter Starlings come from the Baltic countries, including Poland, where Orioles are numerous!

For sheer song quality, we generally regard the Nightingale as Britain's top of the pops—and justly so. One of Northward Hill's most precious assets is its Nightingale population of usually between half a dozen and twenty pairs —quite a fair density. Territories are claimed in the densest scrub, and all of the traditional song areas, first occupied each spring, are centred on patches of impenetrable blackthorn. As with so many of the birds, the recapture in subsequent years of ringed birds confirms the tendency to return to the same territory. Usually Nightingales remain inconspicuous, deep in cover, while they sing—day *and* night, for it seems that only the reduction in volume of competing song, and thus the greater conspicuousness of the Nightingale's, lies behind the idea that they are strictly nocturnal singers. Fortunately one of our birds—ringed as it happens—scorns this secrecy, and for the last three years has taken up his territory alongside the public footpath through the wood, and has selected one of the footpath signposts as his main songpost. In full song, he often allows an approach to within about twenty feet before flitting off, with a flash of chestnut-red tail as he goes, into the thicket to continue his song.

About this time of year we have problems with our good songsters. So that population changes within the wood can be monitored, and the effects of management policies or natural environmental factors assessed, a census is taken each year of the breeding birds. In this census, part of the Common Birds Census organised each year by the B.T.O., the positions of territory-holding birds of the various species are plotted on large-scale maps. Now one of the best ways of locating a territory-holding male is to hear him in song. With birds like the warblers and the Nightingale, which sing from deep cover, it is obviously necessary to be able to identify the various songs without seeing the bird. Plenty of good records exist to allow us to get into practice before the season starts, but they, and the textbooks, *and* the birdsong pundits, forget one thing. Many birds love to mimic each other—indeed this may be how many of them learn to sing at least part of their repertoire. Northward Hill has unusually high densities of Nightingales, Blackcaps and Garden Warblers—so they must all sing in close proximity to each other. Most birdwatchers acknowledge some slight difficulty with the last two—but, they say, it's easy when you get the hang of it. A couple of summers back, we decided that we should check ourselves on this, so on each occasion we identified a Blackcap or Garden Warbler in song, we tried to get a glimpse of the bird itself to confirm the identification—no easy task, but we found that we were wrong as often as we were right! Of course, this could indicate that the four pretty experienced birdwatchers who were involved at one time or another in these checks were not quite as hot on their birdsong as they liked to think! Perhaps so, but perhaps also their suggestions as to the reason for the discrepancy should be taken to heart by others who remain convinced that they *can* tell. There is some aesthetic argument as to the order of merit in which the songs of Blackcap and Garden Warbler should be put, but all agree that the Nightingale is supreme. What we found was that the nearer one species was to another species' territory, the more similar the two songs became, through extended song competitions and mimicry. So a solitary Nightingale in a patch of blackthorn, surrounded by perhaps two or three Blackcaps or Garden Warblers in the oaks, influenced these to such an extent that, although they could still easily be distinguished from Nightingales, they were difficult to separate from one another—you could say that Nightingale tuition had improved each beyond recognition.

Walks through the wood in June give a measure of the activity of the nesting birds—everywhere the adults seem to hasten about their business, and it would be possible to say that some of the Starlings and tits even look a little harassed as they dash backwards and forwards with beaks full of worms or caterpillars. At this time, too, it pays to think just a little about the distribution of the various species, and where the numbers nesting are greatest. Such species- or numbers-rich areas obviously merit conservational interest, effort and expenditure in summer, but are they the same areas that we saw hold, feed and shelter large numbers of birds during the winter? Often not. All too often, the birds of one season (even occasionally a single species) dominate the thinking behind the management of a reserve or sanctuary, but, regrettably expensive though it may be, management must be geared to the year-round values of the reserve.

Looking out over the main clearing, where plant diversity is greatest, the diversity and quantity of available food—berries, seeds, insects and what-have-you—associated with this is reflected in the number and variety of birds nesting. Dunnocks, Robins, Wrens, various thrushes, Nightingales, at least six species of warbler, three of crows and three of pigeons, half a dozen finch and bunting species and perhaps even a Skylark or Corn Bunting contribute to this richness, but to look at it, the clearing is just grass, a few weeds, clumps of blackberry and the occasional hawthorn bush. Yet let this jungle go too far, let the blackberry cover all the grass and ultimately hawthorn will supplant blackberry on the way to climax oak woodland again (a process taking two or three hundred years, admittedly) and all through this maturation, the number of bird species involved will be steadily reduced. This process is one of the laws of nature, and if we value the open scrub stage, we must exert considerable efforts to maintain it against the inexorable pressure that nature will exert. This is why the bills for cutting back the scrub growth each autumn must be met, either in the form of a subscription to the reserve-owning body, or in the form of good, solid, sweat!

One of the more surprising birds, flighting during the early mornings and late afternoons over the open patches, is the Shelduck. In living memory, the wood has not been without a few pairs of Shelduck, nesting in 'burrows' constructed in the tangles of bracken and brambles. These handsome black and white, chestnut-girdled birds probably quite commonly fall prey to the foxes, and all that we find of the nest is a scattering of the distinctive white down and a few eggshells. Those young that do hatch are shepherded by their parents soon afterwards down onto the marsh, and they make their way paddling along the small ditches and on foot out to the major fleets, where they can be reared in safety. This extraordinary passage is put in the shade by the journey (which makes the photographic pages of the local papers most years) undertaken by young reared in rabbit burrows on the North Downs, four or five miles from the River Medway. These ducklings on the way to the river are led by their parents over a Motorway and a major trunk road—many perishing on the former, but usually holding up the traffic on the latter.

On the marshland fringes of the wood, lining the boundary ditch, are some aged, part-hollow pollarded oak trees, and a quiet look at these is well worth-while each June. One, in particular, is favoured by a female Mallard each year —it can hardly be the same female, as I first saw her perhaps twenty years ago— and she builds a nest about six feet up in the hollow crotch. With a cautious, and absolutely silent approach, she usually sits tight and can be carefully inspected at a two-foot range. Occasionally, we have been lucky enough to arrive more or less at the moment of hatching—and we take our look at the rather bedraggled, still-damp, ducklings before leaving quickly to let her marshall them in peace for the jump (at one day old!) down into the ditch.

JULY

By the beginning of the month, many of the young Herons will have fledged, and a scan across the marshes with binoculars will reveal the pale grey, and somewhat ragged, youngsters standing in groups or trying their hands, singly, at fishing in the fleets. Some remain, however, and now that they are well grown, tend to stand on the nest and use it as a practice launch-pad for wing exercises. With the addition of odd branches as ornaments through the season, some of the nests are really huge. I can remember as a boy, when the Herons were in the oaks, and much more easily reached than in their present situation in elms, mastering my fears and clambering up. Strangely, the problem lay not so much in reaching the nest, but in climbing *into* it when you arrived. As big as a dining table, cumbersomely bulky but rather fragile at the edges, they instilled in me a wonder of their size, coupled shortly after with the worry, once in, of how on earth to get out again. The bigger ones were substantial enough to sit in without risk to life, limb or nest.

Many are adequate enough to carry tenants other than Herons. Tree Sparrows are common nesters in the bulky structure, with House Sparrows, Starlings and Jackdaws in a few. In a couple, there are Herons on top, Jackdaws in the middle and Tree Sparrows in the basement!

At this time of year, some of the nests are still with eggs. They are much smaller than these tenements: so small even that the sitting female 'overhangs' at each end, and that when she leaves, the eggs can be clearly seen by looking up through the bottom of the nest, like a Turtle Dove's. These, in all probability are the nests of young birds or newcomers to the colony, nesting for the first time, lacking in authority when it comes to selecting a nest site and, as is common in nature, many will fail—victims of inexperience but part and parcel of evolution.

The parent Herons feed over a very wide area, perhaps forced to do so now because of the agricultural developments on the marsh, the straightening and clearing of the ditches and their replacement in places by piped drains. Some will feed up towards Gravesend, some cross the Thames into Essex, some will use the upcurrent of air created by the slope of Northward Hill to make enough height in slow spirals to overfly the ridge and feed in the marshes of the Medway, or even up to 15 or 20 miles away on the Isle of Sheppey. These last birds, on their return journey, come in over the wood with a great deal of altitude in hand and no means of making a steady descent to the colony. They seem to delight in 'whiffling' in—a breath-taking aerobatic display as they tumble helter skelter out of the sky, twisting from side to side in a series of tight spirals and half-turns—a tremendous contrast to the steady, laboured ascent as they go out to fetch even more food.

When returning from a distance, the parent Heron will swallow into its crop the frogs, eels, water voles, fish and waterbird chicks that it has caught, although those fishing nearby may return swiftly with a writhing eel held crosswise in the beak. Long before the parent has reached the nest, its young will have spotted it and the excitement becomes intense and the noise both indescribable and unbelievable. There is a great deal of agitated jumping around and wing flapping, and the returning bird must take its eyesight, if not its life, in its hands as the young rush at it, grabbing and pecking at its beak to stimulate it to disgorge its cropful of food. This it does into the floor of the nest: strangely, this seems to be the only way, and the youngsters then help themselves. Thus, if a young bird, even quite well grown, strays from the nest and cannot clamber back, it will starve because the necessary stimuli are not there to persuade the parent to feed it. Similarly, young in the flimsy nests we saw earlier may also lose out, as much of the food (some not even dead) will fall or wriggle through the floor!

For the purposes of listening to the music of birdsong, a walk through the wood in July tells you that high summer is here. So far as most of the woodland birds are concerned, the business of territory claim-staking, attracting and keeping a mate demands song of the best quality to be produced at the greatest frequency and, with some exceptions, they have now achieved what there is to be achieved. The young are on the wing, many by now quite independent of their parents and perhaps unlikely to come in contact with them ever again. Both the reduced volume and the leisured quality of the song confirm this suspicion that, for a while at least, the hard work is over. Temperatures on a fine day, with any breeze eliminated by the trees, soar high, and after noon on a fine day the wood may be nearly silent. Mark you, even for us a walk through the clearings in the oaks can hardly be called a pleasure: the bracken by now is over head-high, and whilst you may slide through the vertical stems, every so often a fresh, strong and *very* thorny bramble shoot, running horizontally, brings you to a painful halt.

Periods of restful and quiet observation—preferably in the shade—seem called for, and after some minutes of stillness and silence (very precious commodities for the naturalist) the senses take in more of what is going on. Rustlings in the long grass reveal themselves briefly, and tantalisingly, as shrews (probably pigmy shrews here) bank voles or long-tailed fieldmice (these two separable on colour—the bank vole being a pleasantly rich chestnut brown; and by the round head of the vole, and large ears and long tail of the fieldmouse). Down towards the fringes of the marsh it would be very surprising on a hot day not to see several grass snakes—even one or two quite large ones—perhaps three feet long and thumb-thick—and many lizards scampering around in the shorter, rabbit-cropped grass.

Everywhere birds *are* moving, feeding quietly under cover. Most are still in the drab juvenile plumage, although here and there a 'teenager' is visible —a young Robin, for example, with blood-red patches on its breast beginning to replace the drab brown spots. Even allowing for the obvious fact that so soon after the breeding season youngsters are bound to outnumber their parents very considerably, the number of adult birds to be seen is still surprisingly low.

One reason for this may be that the business of moult, and keeping up the food intake that this energy-sapping business demands, is well under way. Equally, if the adults are in heavy moult, their powers of flight may be reduced—very much so in some cases—and they will tend to keep very much to the safety of thick cover. Feathers are one of the strikingly distinctive features of birds—essential for flight and warmth—and often serving a multitude of other vital functions, used for example in display when attracting a mate. Made largely of the protein keratin, feathers must wear out. For example, some of the female Blue Tits in our boxes have wings about 5 mm shorter (on the standard 60–63 mm) at the end of the breeding season than at Christmas. Naturally, nesting in holes, and the frequent visits, first with nesting material, then with food, wriggling through a tight fit each time, contribute to an unusually high rate of wear, but in all small birds, a year is as much as can be asked of a feather before the need for replacement arises.

These new feathers are secreted by special cells in the surface layers of the bird's body, and the process must take time, and of course must use up chemical energy. Not only is the study of moult fascinating biologically, but it assumes considerable importance in conservation as one of the major periods of stress during the bird's year. Thus each bird we catch in our ringing studies is quickly examined, and any moult is recorded on specially designed cards. When sufficient information has been accumulated, we can analyse it to see how long the difficult period is, when it occurs, and how it is influenced by the weather and so on. The results are fascinating.

The Blue Tits and Great Tits—residents year-round in the wood—start moulting at the time (or sometimes before) the young leave the nest, in early June. From then on, moult proceeds at a leisurely pace. Rarely is more than one primary flight feather completely missing from each wing, although others will be in various stages of growth, and rarely will the 'missing' primary feather-area exceed 20% of the total. Thus it would be surprising if the bird were ever inconvenienced, or endangered, although there are obviously slightly greater risks—for example of being caught by a predator. Food demands are not much increased by energy needs of this level, and when moult occupies nearly one third of the year, as in the tits, few problems are likely to arise. This relatively slow moult seems to be general for our resident species, but (as always in nature!) with exceptions—and these are difficult to understand. Two examples are the Dunnock and the Linnet, which moult in fifty or sixty days (perhaps sometimes less) against the hundred or so days of the tits. Perhaps food requirements as yet unknown dictate this speed, or perhaps it is a throwback to times past when these birds migrated. (Some of our Linnets, in some years, do move down to southern France and to Spain, and the Dunnock on the continent is a regular migrant—as are several other species that we in Britain consider to be sedentary residents.)

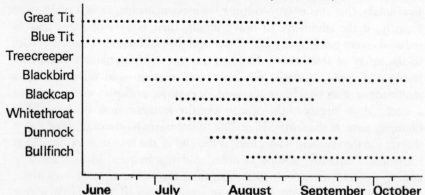

MOULT SEASONS—the dotted lines indicate the extent of the period that the various species are in moult. Note how the resident birds tend to take longer (the Dunnock is an exception) and how quickly the Whitethroat completes the process to get away south on migration.

The problems for a migrant bird are *very* different. Having raised one or two families, they must move south before the food supply begins to fail and the climate to change for the worse. Good examples of these pressures are demonstrated in two species that we see most of in July and early August, the Whitethroat and its less conspicuous relative, the Lesser Whitethroat. Having finished raising their last broods in the clearings or in the scrub patches early in July, most of the Northward Hill residents will be on the move southwards towards Africa in the latter part of August. Thus the process that takes a similar-sized bird, the Great Tit, a hundred days has to be rushed through in about forty-five by these two whitethroats! On occasions, we have caught Whitethroats with six of the ten flight feathers missing (and perhaps with only about 20% or less of their usual flight capability—enough to cause wonder at how they managed to get far enough off the ground to get into the net!). Not unnaturally, when in such heavy moult these species skulk even more than usual—not only, we suspect, because of the difficulties they have in flying, but because they must eat prodigiously to supply the necessary energy for new feather production. The Blackcap has an only slightly less rushed moult, but its close relative, the Garden Warbler, has gone about things another way in the course of evolution. Garden Warblers only extremely rarely moult in this country—usually they delay moult until they reach their wintering grounds in Africa.

Young birds face different problems: there is no need for them to change the wing feathers they have grown in the nest—they will hardly have had a chance to wear out yet. But, and it is a big but, when in the nest, and shortly after fledging, they need to be as inconspicuous as possible to protect themselves, by cryptic camouflage, from predators: hence the drab browns and greys, the streaks and the spots. Long before next spring, however, they will be wanting to compete with their elders for territory and for mates, and for this they need all the paraphernalia of full plumage for display purposes. Thus they need to moult their body feathers alone, and in July and August we see large numbers of the rather ragged, parti-coloured teenagers like the Robin we encountered earlier.

AUGUST

Even the halcyon days of high summer bring us some surprises in the wood. When for many years you have spent much of your time birdwatching regularly in one area, you not only get the feel of the place and become familiar with its regular inhabitants, but you get used to the surprises that it can occasionally spring. For us in Northward Hill unusual birds are almost always in the peak passage months like September and October. I suppose that if we accept that the dispersal of the young resident (not migrant) birds may start soon after the breeding season, we should expect strangers in August as well—and we do get them.

In summer, water is at a premium. There are only a handful of ponds in the whole 130 acres, none of them larger than a small room, and in a dry year only two of them still contain water by August. Of course the vicinity of these ponds make magnificent netting sites, for sooner or later we will intercept most of our birds on their way to or from the water. I can remember a very hot, but fairly quiet day so far as bird movement was concerned, a couple of Augusts back, when we were having a leisurely lunch which was interrupted by the sharp—and quite unmistakeable—call of a Kingfisher. There are none breeding on the marsh, and very few in the whole of north Kent, and they are not the sort of bird you expect to find in mature woodland anyway! We were arguing the toss about the likelihood of a mistake when the bird flew by, settling the matter. How it located the pond, deep in the trees, I shall never know as it was flying characteristically low—only a few feet above the ground. Find it it did, however, and a while later, after it had been flying around a little, we caught it and ringed it. This is unquestionably a moment of privilege for a bird ringer—beautiful Kingfishers are at a distance, but how much more so when you can turn the bird in your hand to see how the optical effects of the flashing turquoise-blue back feathers work. Now you can see the astonishingly tiny coral-red legs and feet—all out of proportion to the head end with its very workmanlike beak. After we released it, the bird moved on, and a couple of weeks later we heard the sad news that it had been killed, flying into a window in a riverside house in the Cotswolds that had claimed other Kingfishers before.

We have to thank our regular netting for research purposes for our only Marsh Warbler record, too. This species, except when in song, is nearly impossible to separate from the very much commoner Reed Warbler other than in the hand—and even then, the differences are subtle enough to be a matter of statistics. Surely the Marsh Warblers must have an easier way of telling? So close to the marshes, we expect to encounter a few Reed Warblers actually within the confines of the wood and, indeed, one we ringed was killed in Portugal later in the autumn, but this was an unexpected bonus.

More regular as August birds—all over the wood—are Hawfinches, but their regularity detracts little from their fascination. Most years a few pairs breed—very discreet they are too, and most difficult to locate until the young are just about on the wing. Soon after fledging the family parties move out into the orchards to take advantage of first the cherries and later the plum crop —obviously not nowadays at a level likely to annoy the grower, although earlier this century this was the case in some parts of Kent. Would that I could have seen it! In August, they return to the fold because the blackthorn fruit—the sloes—are ripe, and we see much more of them again. In the hand, the full power of their massive beak—with an equally massive skull to which the necessary jaw muscles are attached—can be seen. The silvery beak is pyramidal, and about $\frac{3}{4}$ in long, wide and deep. Some well-equipped research ornithologist has measured the pressure that the bill can exert at over 180 pounds per square inch! Such huge pressures are obviously needed to break open cherry and damson stones to reach the highly nutritious kernel within, and the Hawfinches' other main food, the seeds of the hornbeam tree, also are very tough to open. Never having had the equipment to measure this pressure accurately, I can only testify to the fact that being bitten by a Hawfinch hurts more than somewhat—and you do not pull your finger away, else a triangular piece of you remains with the bird. Watching another ringer wait for the Hawfinch to let go of him is a much more attractive prospect.

In the hand, the striking white wingbar—so conspicuous and allowing you to see 'right through the wing' in the field—is less significant against the purple sheen, and the extraordinary 'fringed' ends of the secondaries—the inner flight feathers. It would be most interesting to know what these feather tips, splayed like a Kaiser's moustache, do to the aerodynamics of the wing, but it has been suggested that they make an unusual noise which is part of the Hawfinch display flight.

August sees the fruiting of most of the plants that carry berries, both in the clearings and within the trees. Primarily this involves the blackberry and the elder—and as any gardener knows, the birds often do *not* bother to wait for the fruit to ripen before eating it. For many species, like the warblers, about to migrate southwards, this sugar-rich fruit is absolutely vital. It can be eaten quickly, needs little searching out, and can be obtained in prodigious quantities with little effort. What is most difficult for us to understand is how a bird like a Whitethroat or a Blackcap, having fed for so many months almost entirely on insects, and having raised its young on the same diet, can suddenly change from this difficult to digest protein-rich diet to one so completely vegetarian.

It seems most probable that this dietary change occurs not long after the last brood of young have fledged, and at the start of the rapid moult that we were talking about earlier. If this is the case, then doubtless the high-energy food helps cope with the strain of the moult. But there is another very significant purpose. Stored fatty materials or lipids serve as an insulating layer and as an overnight energy store for our resident birds, especially during the cold winter weather. For the migrants, similar fat stores serve as fuel, to be used up steadily *en route*. The greater the quantity of stored fat, the greater the non-stop range of the migrant.

Recent studies of bird weights, both in the U.S.A. and in this country (where Northward Hill birds have played their part) continue to reveal astonishing aspects of bird migration. How birds navigate remains perhaps the greatest unsolved mystery in biology: now we know that not only do they get there accurately, but some reach Africa in one hop, and it is conceivable that others even overfly the Sahara from southern England. Probably the best illustrative examples are the Reed and Sedge Warblers: the former only increases in weight by about 30% in autumn, and ringing recoveries (like the one I mentioned earlier) show us that they make landfall, after leaving this country, in southern France and the Iberian peninsula—just about the distance the energetics experts tell us they should be able to make. Here they rest and feed furiously, perhaps for several days, laying down sufficient reserves for the next stage of their journey south. The Sedge Warbler, on the other hand, increases in weight by 100% or more—layers of fat everywhere, the bird nearly spherical, and certainly having to struggle to get airborne—and we have almost no ringing recoveries between here and their winter quarters in Africa, often south of the Sahara!

The graph (page 126) shows the rate of weight increase in autumn for the

Blackcaps in the wood. These, as you can see, are not in the great rush to leave that the Whitethroats are. The Whitethroats, in that month between finishing the breeding season and leaving for winter quarters, must moult *and* somehow eat enough to store about a 30 to 40% addition to their body weight (they, too, are several-stage migrants)—in terms of human comparisons, an effort to turn Billy Bunter pale.

The Little Owl family have at long last fledged from their nest, which is situated in an old, stubby, lightning-blasted oak (as in most years). The youngsters are painfully conspicuous during the day, and this year the three are each in different trees, and each is surrounded by a frenzied gathering of scolding small birds. This furore makes them easy for us to locate, and as we approach it is only too obvious that they still have much to learn in the way of fear of man, for they sit, glaring boldly at us with bright yellow eyes, 'clicking' in an irascible manner and bobbing their heads up and down. The bits and pieces of fluffy down still sticking to their heads and facial discs give this stern demonstration of annoyance more of an air of farce! Each year when the young have left the nest we check through the contents (by now revoltingly smelly) for although they feed largely on worms and beetles, they do take birds, and we might find the mortal remains of some of our ringed population. This year we have struck it rich: two Starlings, a Song Thrush and, most surprisingly, a Mistle Thrush, and an adult at that. There cannot be that much of a size difference between the Little Owl and the Mistle Thrush —and Mistle Thrushes are amongst the boldest when it comes to seeing off marauding Magpies, Jays and Hawks. Unless the thrush was taken by surprise, that must have been a battle royal. Little Owls tend to hunt by sitting motionless on a branch overlooking a more or less open area, and when prey passes beneath them they 'parachute' down on top of it, striking with their talons— it may be that this was the way of the Mistle Thrush's end.

SEPTEMBER

Before the leaves begin to yellow and fall with the onset of autumn, we may take stock—on a walk covering the whole wood—of the toll taken during the summer by Dutch Elm Disease. In much of Britain, various strains of the disease have persisted since the last major outbreak before the war, resulting in the odd branch yellowing one summer, and remaining stag's-antler-like for some years before dropping off. In part the disease may have been responsible for the evil reputation that elms have for shedding limbs without provocation.

Small wood-boring beetles—related to the domestic furniture beetle—spread the disease, which is primarily caused by a fungus that blocks the water carrying vessels of the tree—causing initially a leaf yellowing—as when the leaf petiole vessels are closed naturally each autumn. The recent series of mild winters has encouraged an increasing beetle population as the grubs beneath the bark have not suffered the usual mortality due to frost. Coupled with this, and as is common in diseases both of animals and plants, steady evolutionary pressures have led to the development of more virulent strains of the fungus—and this combination has given rise to the present outbreak of the disease which is devastating the elms of this country.

The first signs appeared in Northward Hill in mid-summer 1969—and from the speed with which it reduced a patch of elm trees about the size of a tennis court from full health to leafless skeletons it was at once apparent that the disease was severe. When I left for a field trip to the seabird island of

St Kilda, all was well. Twenty days later, on my return, whole trees—not just branches—in the patch were yellowed, and on some the leaves had already begun to shrivel—just as if a bonfire had been lighted beneath them. Each year since then we have had mild winters, and each year the spread has accelerated until this last summer, when very few elms remained alive to be infected. There is now hardly a live elm, or even elm sucker, in the whole Grain peninsula—and much the same is true of large areas of Kent and Essex, and now other parts of Britain, especially some in the Severn Valley. Should the spread continue, it bodes ill for those parts of England where elm is the dominant tree, especially where it is the principal component of hedges: as dead trees fall or are removed, we shall need an unusually concentrated effort to prevent these hedgerows vanishing entirely—perhaps the key factor will be the cost of replacement. Our main hope must remain that the elms themselves will manage to survive—perhaps only as resistant suckers here and there—for in the same way that the Government has had to withdraw from its position of ordering infected trees to be felled (because of the immediate financial hardship to individuals and the gigantic overall cost of felling literally more than a million trees), it is difficult to see a source of the finance needed for such a huge replacement job.

The elm component of Northward Hill is relatively large: scenically, and aesthetically, they will be missed. In an ecological sense, too, they are most important. The huge leaf area of a mature tree supports an incredible insect population right through the summer, and the cracks in the bark shelter the countless insect (and other arthropod) eggs, or the overwintering larval or pupal stages, that enable this population to reappear next spring. The surplus of this productivity (and it must only be the surplus, or else the predators would eat themselves out of house and home) serves as part of the year-round food supply for countless other animals, including birds, either directly or as links in a food chain.

When a tree dies, falls and decays naturally, the composition of this live-stock changes. There are no longer any leaves, so the sap-sucking insects are not present. On the other hand, the wood becomes much softer, and there are saprobic fungi to feed on, so the burrowing larvae of other insects increase—and support some of the same birds, while others find insufficient to eat and move away, to be replaced by the specialists in finding food in these circum-stances. It is most instructive to find a comfortable seat on an old branch or something, sheltered from the elements and with a screen of vegetation partly shielding us from view, and to watch over one of these fallen trees to see just what comes along.

First of all, we see the woodpeckers: the Great Spotted can chisel into very thick bark and quite sound timber to reach grubs, but normally, the sparrow-sized Lesser Spotted (with its high-pitched hee-hee-hee call) must concentrate on the smaller branches high up, though as the bark rots, even the Lesser Spotted can get to work on the trunk. Other primary feeders include the Magpie and Jay—and they are one of the main reasons for our watching from concealment, for they are amongst the most alert, sharp-sighted and cautious of birds. (This may partly account for their success, and certainly is one reason why so few good photographs of the crow family are ever taken). The technique of these two is less subtle than the woodpeckers': using their bodily strength as well as their beaks, they peel back sheets of bark to get at the grubs, spiders and earwigs sheltering beneath. Squirrels set about the job in much the same way, for they are by no means averse to adding a little animal protein to their largely vegetarian diet (either in winter, or with birds' eggs in summer). None of these search each piece of bark at all thoroughly, and this is fortunate for the army of 'clearers-up' that come behind. Given an opportunity like this, the smaller Blue, Coal and Willow Tits—usually feeding in the higher branches of the trees—join their larger, and stouter-beaked relative, the Great Tit, which often feeds at ground level. Our fallen trunk is soon alive with tits, all calling excitedly, and the racket may encourage the smallest scavengers of all to come and join in the feast, Treecreepers, Goldcrests and Wrens. So there is food for all, and the messy feeding habits of the larger birds are most helpful to the smaller, weaker ones.

Even the tit flock does not carry out a persistent and thorough search—and surprisingly soon the party has moved noisily on its way, leaving perhaps a Wren or a loudly ticking Robin in sole possession. It may be that food resources are better conserved by this 'little and often' form of exploitation—obviously there are some benefits in spreading your larder around, especially when timber merchants are about, dragging off that which they require and burning the remainder!

Perhaps the greatest impact of the disease at Northward Hill was that after only three years, almost all of the heronry trees had succumbed. To our surprise, the herons raised no objections to nesting in naked trees even late in the season (perhaps we should not have been taken aback, as the season *always starts* with the trees leafless) and in the last couple of years only a handful of nests have been in live trees. The decision has been taken, rightly or wrongly, to fell all the dead elms in Northward Hill. An attempt has been made to minimise the impact on the heronry, and elsewhere, by spreading the felling over a period of four or five years, in the hopes that the herons will gradually return to the oaks—where, you may remember, they originated. What the impact on the heronry will be, whether the decision to fell at all was correct, and whether the overall effect on the ecology of the wood will be as dramatic, and damaging, as some consider, only time will tell.

September is *the* one month when binoculars must *not* be forgotten. Over the years we have come to appreciate that Northward Hill is well situated on a migration route, and as one of the few reasonably-sized areas of woodland visible on the easterly aspect of the south Essex and north Kent coasts, it receives its share of migrants—including rarities—at the same time as do more famous spots like Cley, Gibraltar Point and Spurn, where 'falls' of migrants are large and regular enough to merit the positioning of part of the network of coastal bird observatories. Except at Dungeness, the Pied Flycatcher is a rarity in Kent on migration and it does not breed here. Yet each year late in August and in September, a walk through the oaks is likely to give a fleeting glimpse or a brief hearing of the quiet call—and sure enough, later in the day, in amongst the tits in the nets at the pond will be a Pied Flycatcher—and on one occasion we caught two at once. Each year the total slowly mounts to six or seven. In the hand they are very placid birds—a great contrast to the tits—with astonishingly short legs and the disproportionately large wings you would expect on a bird that depends for its meals on its ability to outfly insects. Another bird that we usually associate more with the coastal observatories is the Icterine Warbler, but one September we tracked a large grey warbler from bush to bramble across the clearing before getting a clear view of its bluish legs, long stout beak and long wings, and confirming its identity. Imagine just how many similar occurrences must go unrecorded all over the country—partly by chance and partly because the local birdwatchers are elsewhere looking for the same sort of bird!

One of the September oddities that puzzles us greatly is the Nuthatch—for there have only been two records of this bird in Northward Hill in the last twenty years. To human eyes the wood looks ideal, and indeed, only ten miles away Nuthatches flourish in apparently identical surroundings.

To human eyes anything but ideal for the species, the pond one year attracted a Grey Wagtail. As with the Kingfisher, how the bird *saw* the pond as it flew over—the pond was table-top sized at the time—is difficult to understand, and it is difficult to imagine a Grey Wagtail flying, or more likely hopping, down through the branches to a leafy, very smelly and stagnant puddle covered in mosquitoes. Nonetheless it did—and walked up and down, tail wagging, calling loudly, in all its lemon yellow and blue-grey finery, on the dam that we built to try to deepen the water accumulated in winter so that summer evaporation takes less of a toll.

Another astonishing bird, doing an even more astonishing thing, on the dam at the pond a few years ago was a Common Buzzard. Whilst these are rare birds in Kent, we do expect to see the odd bird each autumn in the wood, but *soaring* high above us on the upcurrent that the ridge creates, not standing on the dam calmly eating an unfortunate Blue Tit youngster in one of our nets. We disturbed him just after he had decapitated the Blue Tit, and he took off straight into the net. Our nets are not designed for birds of this size, but at this take off speed he did not have a chance either to fall into a pocket, or as we feared he might, punch a hole right through. After some ineffectual attempts to push through the net, the Buzzard turned and flew off in the other direction—this time fair and square into a net, and only a step away from us. When he was released shortly after, he carried the usual ring as a memento. If it was difficult for a Grey Wagtail to penetrate the leafy canopy to get down to the water, how much more so for the Buzzard with its three or four foot wingspan!

While we regularly see harriers out over the marsh, especially in winter, and have Kestrels about the wood much of the time, other birds of prey are very scarce. Just after the war, and until the early 1950s, the marsh fringe of the wood was on the regular beat of a Peregrine: then the pesticide-induced population crash dramatically reduced Peregrine numbers, driving the survivors back into their strongholds in northern England and Scotland. The Hobby—our smallest falcon—has probably never been particularly common in Britain—we are right on the fringe of its range, where climate can be limiting, especially as it is an insect-eater. Occasionally in May, August or September, one on passage will pause long enough in the wood to catch a few of our large dragonflies before moving on and we may be lucky enough to catch a glimpse of this handsome grey-backed, red-trousered bird. Occasionally a Sparrowhawk will spend a while in the wood—sometimes even staying over winter and keeping itself well-fed on the roosting birds. So inconspicuous can they be that we will find more remains of kills (with characteristic 'snipped-off' wing and tail feathers) than we will have sightings of the birds.

Such views that we do have will be of birds flying, agilely twisting through the trees at speed and soon lost to sight, but the same cannot be said of a Goshawk that stayed briefly with us some years ago. She (the females are really huge and very distinctive) used to hunt across the clearing, and occasionally soar above it, so we had plenty of time to study her short, very broad and blunt wings, the very fine barring on her underparts and her eyestripes—thick and white, meeting on the nape of her neck. This sight would fire the imagination of any birdwatcher—of all the birds I have ever seen, the purposeful hunting of a large female Gos conveys to me the greatest impression of power—I am content just to stand and drink it in.

OCTOBER

Only the last stragglers of our summer visitors now remain: our ringing results indicate that the few late Chiffchaffs and Blackcaps that we catch—all very high in weight, and with visible white fatty patches under the skin if you blow the feathers apart to look—are late birds from elsewhere (possibly not even Britain) hastening southwards. We will often have several days of 'Indian summer' during the month, and the abundance of insect life to be seen indicates that it is not a shortage of insect food that has driven the migrants away. The Swallows and House Martins from the village—two of the last birds to leave us—are often over the more open areas during the month, cashing in on what the early departers have left uneaten.

Perhaps it is just as well that this abundance of food remains, for October sees the start of an inward rush of birds to replace those off and away south. A glance at a weather map of the northern hemisphere in any atlas will give us some clues as to why. Look at the pattern of isotherms—the lines joining places of equal temperature at a given time of year. Most often, the two maps given are for January and July—both mid-season for the birdwatcher. Some detailed studies have been carried out over Europe to see how bird movements are related to temperature—the classical example being the northward move-ment of Swallows in spring. Birdwatchers all over Britain and the continent—spanning north/south as well as east/west—recorded the date of the arrival of their first Swallow in spring, and the 'front'—the vanguard of the hordes of Swallows moving north—was found to be almost precisely the same shape, and to move at the same speed, as the 48°F isotherm. This brief digression I

think illustrates how close the association can be: now let us look at the January
—mid-winter—picture. Imagine the isotherms in much the same way as you
picture contour lines on a map: in central and eastern Siberia, there is an
unbelievable centre of low temperature, with January averages way below
$-30°C$. The January isotherms do not follow the parallels of latitude, so that
latitude for latitude, the oceanic coast of western Europe is very much warmer:
Kent may be up to $30°C$ warmer than Lake Baikal on the same latitude in
Russia. For migratory birds wishing to take advantage of the summer climate
and food supply in the vast areas of northern Europe and Asia, a considerable
barrier exists to the simple north/south migration that might seem obvious.
It is formed by the belt of mountains from the Himalayas right across to the
Alps north of the Mediterranean. Whilst these mountain ranges are not
impassable, they have lent their weight to the development of a marked south-
westerly trend in many bird migration routes, from central Eurasia across to
eastern Europe. Birds move away from climatic conditions which make it
impossible for the majority to survive (there are, naturally, specialist exceptions like our Ptarmigan that can continue despite the snows because of
adaptations in both structure and way of life) and for many, the most practicable route—defined by millenia of evolutionary bungles and successes—
has become the southwestward one to Britain and Ireland. Thus we, on the
west coast of Europe, must expect perhaps even an increase in the total number
of birds that our countryside supports when winter comes.

An additional factor may also influence the numbers of some of our winter
visitors. A series of good breeding seasons and mild winters, with an adequate
food supply, may allow their numbers to build up to such an extent that their
sheer weight imposes impossible strains on the available food, and possibly
also on the social organisation of the species concerned. The best known
example of this phenomenon in animals is the approximately four-year cycle
in Lemming numbers. When numbers of these attractive little rodents—
widespread in northern continental Europe—reach their peak, the well-known
westward mass emigration begins in search of pastures new. Often the size
of the migration, the obviously abnormal mental state of the lemmings, and
the hordes that perish in fruitless attempts to cross the Atlantic Ocean will
make newspaper headlines.

It seems likely that much the same biology lies behind the periodic arrival
in this country of some most attractive birds—the so called 'irruptive' species,
the Crossbill, the Waxwing, and (less commonly) the Nutcracker. Because
these are birds that we rarely see, and relish when we do encounter them, I
think we tend to overlook the possibilities that other species that we commonly
see may behave in the same way. The tits are a good example. In much of
northern Europe they needs must be migratory anyway, and when numbers

are built up by a long period of good conditions, we may see an absolute avalanche of birds into this country. The last time this happened, in 1957, we saw not only extraordinary numbers, but some extraordinary behaviour. No milk bottle was safe, and there were many complaints in letters to the newspapers about tits entering houses and tearing off strips of wallpaper! Unless we soon have a severe winter, it seems fair to predict that the time is becoming ripe for a repetition of this behaviour, and the last two autumns have seen mini-happenings on the continent that may presage this.

Walking round the wood in October, you are conscious of just how many tits there are. Some of the huge parties have not yet broken up into their winter territories, and catches can be considerable—perhaps even a hundred birds at a time, which keeps us *very* busy. An inspection of the nets is impeded by the numbers of falling leaves that also collect in them—the bane of a mist-netter's life. Usually we can tell any 'foreign invader' Blue Tits more or less at a glance, for they seem to be very drab plumaged birds with a distinctly pale grey appearance.

We can visualise the pressures building up on the birds now—food supplies in some cases are diminishing and alternative sources must be found, in competition with the more experienced adult birds that have faced the problems before. Some research elsewhere has suggested that by November most of the young that are going to succumb will have done so. Results so far suggest that this is probably not the case in Northward Hill, but before next spring, die most of them must. This seems a brutally matter-of-fact statement, but no old adage is truer than 'nature red in tooth and claw'. Consider the situation for the tits—arithmetically and biologically it is relatively simple. At the start of the breeding season, there are two birds, male and female. At the end of the season, the average pair will be the doubtless proud but harassed possessors of a family of ten youngsters: the family has suddenly grown to twelve in all. A similar situation persists for other species, of course, and we must remember that we are dealing (very roughly) in averages, not just considering the most fruitful unions! Unless the population of Blue Tits all over Britain is to escalate frighteningly, a threat to its own food resources and those of other species, this situation must change. With no mortality operating, and breeding continuing at this rate, our one pair of Blue Tits will have given rise to more than 2,000 individuals after five years. Thus our twelve birds must be reduced, by the start of the next breeding season, down to two again if a steady state is to be maintained. Speaking approximately again, this means that probably one of the parents will perish overwinter, and nine—90%—of the young. Is all this death necessary, though? Evolutionarily speaking, the answer must be yes, because in this way the population retains sufficient elasticity to exploit environmental changes or to cope with adversities like severe winters or epidemic disease: if the clutch were to be tailored precisely to size, there would be no leeway, insufficient flexibility in the system to allow the situation to be quickly remedied.

It is quite fair to say that, while we can make this sort of calculation and sweeping statement about mortality, it is much more difficult to assess how the recovery phase works. The Common Birds Census mentioned earlier showed the dramatic effects of the 1962–63 winter on woodland birds, with population reductions in various species of between 20% and 80%. In Northward Hill, as elsewhere, most were back to 'normal' levels by 1966—one or two earlier, and one or two (like the Green Woodpecker) are still on the way back in the 1970s! Yet an analysis of the breeding results in the 1963 and 1964 summers showed no real evidence of the increase in clutch size that many people expected. Despite this, the populations recovered. In some way, presumably, the mortality rate had been reduced, allowing numbers to increase again. Patently it is ridiculous to suggest that cats had declared a truce with Blackbirds and Song Thrushes, or that car drivers were taking extra care as they knew that numbers were low. The explanation of this process is still being sought, and until the reason is found, biologists and birdwatchers together will be able to delight in arguing about, for example, this apparent evidence that mortality rates are influenced favourably by low population density.

Looking back to earlier months, in the autumn of 1968 we 'said goodbye' to our usual population of about twenty pairs of Whitethroats and their young of the year. In 1969, only a single bird (we have no evidence of a pair having bred) returned. Birdwatchers, censusers, ringers and bird observatories all over Britain in the spring of 1969 were immediately in touch with the B.T.O. headquarters in Tring, all with the same question that we had to ask: 'where are all our Whitethroats?' The answer proved far more difficult to find than we had imagined: obviously the disaster (accounting for about 80% of the nation's Whitethroats) had taken place overseas—either on passage or in the unknown winter quarters of our birds in Africa. Many possibilities were checked—the weather in the Mediterranean area, the use of anti-locust chemicals (there was a locust plague that winter) and so on. Only recently has a possible answer come to light: a reversal of the short winter rains in central Africa may have resulted in a more or less complete lack of food for the migrants to fuel-up on for the northward journey in early spring. Birds caught in Africa were so lightweight that the ringer handling them felt they had no chance of flying for long enough to cross the Sahara. The few Whitethroats that survived have had a much harder time returning to numbers than did birds after the 1962–63 cold winter: three breeding seasons later, Whitethroat numbers had improved only slightly, but in 1972 the handful of pairs in the scrub areas again gave Northward Hill the true sound of summer.

NOVEMBER

In October and November, the wood begins to swarm (there is no other word for it) with Goldcrests. These tiny little birds, at about 5 grams (less than one fifth of an ounce) the smallest of British birds, have in the last couple of years summered and bred in the wood, but the autumn influx of some hundreds must come from elsewhere. They are always on the move, wings flickering, and piping their tiny shrill call. From twig to twig, bud to bud they hop—first upright, then upside down. I have watched many times these feeding movements—often at very close quarters—too close to use binoculars, and as yet I cannot answer the question of how they land upside down. Do they fly alongside right-way-up, and then do a sideways roll to perch feet upwards, or do they approach at low level, and then 'loop the loop', perching at the top of the loop? We assume that at this time the bulk of their food is the eggs that various insects have secreted in the cracks in the bark—the Goldcrests use their needle-fine beaks to get at these. At close quarters, the crest on an actively feeding bird is a thin, dark-bordered, pale yellow line down the centre of the crown, but should one male trespass too closely on the feeding space demanded by another male, the head feathers are fluffed out and the crest raised as in the courtship display. Then the amount of yellow feathering can be seen, and the deep flame bases to the feathers show, startlingly, to best effect.

It is always difficult to grasp the fact that these tiny, fragile birds have reached this country after crossing the North Sea. One would suppose, surely, that something so weak and fragile just could not survive the length of the journey, let alone the buffeting. The first indications we had that the Northward Hill Goldcrests originated overseas came with two recaptures of ringed birds: one, ringed on arrival in a 'fall' on the north Norfolk coast, was caught by us eight days later, and two other birds ringed in Lincolnshire were re-trapped by us in the subsequent autumn. Then came the greatest surprise of all. It will be apparent that many, and sometimes most, of the birds we catch have already got rings, and our own rings at that, but nevertheless we usually check the ring number as we remove the bird from the net—just in case! On this occasion the whoop of interest and joy that accompanies the discovery of a bird from elsewhere was followed by the question 'where is Hiddensee?' On our return home we discovered that Goldcrest Hiddensee No. 90254961 had been ringed under the East German scheme on the little island of Hiddensee in the Baltic a fortnight earlier. We know from its retrap history that this bird stayed on with us right through the winter.

Occasionally, the Goldcrest flocks are accompanied by a similar-sized, and even more handsome relative, the Firecrest. There is little difference in the colour of the crest—despite what the names suggest, the females of both species have yellow crests, the males of both, flame-coloured with a greater or lesser degree of yellow tips. Otherwise, the Firecrest can be quickly distinguished by the black and white horizontal stripes across its face above and below the eye, and by the golden bronze mantle—much more striking than the Goldcrest's plainer olive. Perhaps it is because they are such small birds, but the mantle feathers have a superfine quality all their own.

November is the time that the tit flocks contain the greatest variety—the huge parties of August, September and October were overwhelmingly composed of Blue and Great Tits, and the very small flocks of January and February are predominantly these two species. In November, we are in a no-man's-land, when Coal Tits are frequent—sometimes some continental birds with their very much richer colouring are recognisable amongst them. Whole flocks of Long-tailed Tits may appear in the nets—these delightful birds have very strong family ties, and call incessantly to keep in contact with their relatives. Should we catch one, the likelihood is that we will catch most of the party, but if not, the others will wait, calling high in the trees, for the trapped birds to be ringed and released—when all will go on their way together. Again these birds are surprisingly mobile considering their wings and disproportionately long tails, and parties of ringed birds regularly commute between High Halstow and the nearest ringing station, at Cliffe, a few miles away.

Before the Long-eared Owls ate the last pair, we used to catch the Marsh

Tits very regularly, but now we only catch their close, and difficult to distinguish relatives, the Willow Tits. Many field guides dwell long on the problems of field identification of these two species: if your bird calls, things are rather easier, for the 'pit-chu' of the Marsh is quite distinct from the 'zeee' or 'chay' of the Willow—but you may not be so lucky. While there is sometimes a difference to be seen between the glossy crown of the Marsh and the drab one of the Willow, feather wear, damp, or lighting conditions can alter all this. Similarly, feather wear can soon remove the pale feather fringes that cause the pale patch on the Willow Tit's wing. When we have the two species in the hand, one other difference is apparent: the Willow Tit is much more bull-necked. There doesn't seem to be any nape at all, and the black crown finishes in the middle of the back. The biological explanation behind this difference is a fascinating one: the Marsh Tit Nests in natural holes and crevices, while the Willow Tit excavates its own nest hole in soft or rotten timber—and the bull neck houses the muscles that do the excavating. This difference requires just a little practice before it can be reliably used in the field!

Early in the year we saw the importance of subcutaneous and body fat to thrushes when nights were cold and long, and when food was short. It is at about this time that Blackbirds, for example, begin to store fat—resulting in weight increases. The graphs show the year-round picture for several species.

With the tremendous influx of thrushes and Starlings to the wood at this time of year, it is really no surprise that on successive walks the crop of hawthorn berries can be seen to be steadily diminishing. Partly this is due to the weight of numbers, but partly I suspect it is because of an increased appetite for sugary things—a parellel to the warblers and the berries at the end of the summer—and the winter weight increase of the Blackbird is based on these berries. Superficially, we could I suppose argue that the thrushes are running a severe risk of eliminating their food supply when they don't really need to:

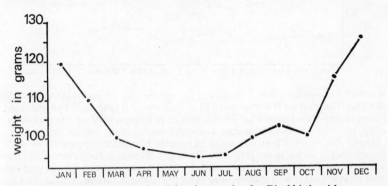

WEIGHT I—classic pattern of weight changes in the Blackbird, with 20 or 30 grams of fat deposited to tide birds over cold nights and extended cold spells.

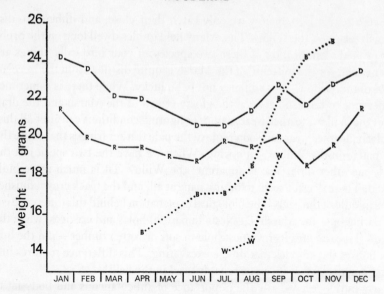

WEIGHT 11—average patterns of weight changes through the year for four species. Note similar patterns of the two residents: Dunnock (D) and Robin (R). Each puts on some 10% of its normal body weight to act as a winter 'insurance'. The two migrants, Blackcap (B) and Whitethroat (W) put on weight quickly before leaving in the autumn. The Blackcap, leaving later, probably journeys south in fewer but longer 'hops' than the Whitethroat, hence the two-thirds increase in weight, perhaps mainly fat laid down for migration 'fuel'.

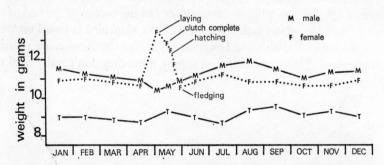

WEIGHT 111—pattern of weight changes through the year for male (M) and female (F) Blue Tits, and for Treecreepers (T) of both sexes. It is thought that the rapid rise in weight in the female Blue Tit just before egg laying is partly due to the storage of fat, partly to the obvious increase in size and activity of the ovaries, and partly to changed diet which allows this rapid increase to take place. Apparently no such dietary change occurs in the male Blue Tit, or in either sex of the Treecreeper. The weight of the Treecreeper shows remarkably little fluctuation through the year—as little as in any other bird species—and poses interesting questions. Why is there no breeding-season increase in the females, and how do Treecreepers survive winter weather so well, with (apparently) so little protective fat?

but perhaps not, perhaps the essential is to be in the right condition to face the problems of winter. If a bird starts off with suitable reserves, a minimal daily food intake will suffice to keep it alive, whereas if it waits until daily food, fat reserves and fatty insulation layers are all needed *at the same time*, the chances of successfully meeting these demands from available food are poor, especially in the available time.

Should the weather on the continent be poor, November days are regularly punctuated by the sight of flocks of Chaffinches, Skylarks and Starlings moving westwards up the Thames. The roosts are back in full swing—and indeed many birds just do not need to leave the wood during the day. Plentiful food is available, and new arrivals like Redwings, with their thin 'seep' call, or Fieldfares, with their harsh chuckle, can be observed at close quarters, for many of them have as yet little familiarity with man. Surely these two must be amongst the most handsome of our birds—especially the Fieldfare with the rich subtlety of its greys, chestnuts and blacks.

DECEMBER

Winter has returned, sometimes in earnest. The days are short, grey and dark, but until the end of the month temperatures cannot be described as really low. Roughly two years in three sees a white Christmas in Northward Hill—surely the ideal time for the tracker to interpret what is happening. We can follow the bold tracks of a pheasant, punched through the thin crust on the snow, and identify it as a male when we find the tell-tale thin line where the long tail has been allowed to drag in the snow. The spots where it has stopped to forage are only too obvious—a black-brown scar where a bucketful of leaf litter and snow has been churned up. Sooner or later, surely, it must take off—for the impression of the wing feathers in the snow, when we do find the spot, shows more clearly than any other way the lift that the apparently ineffectual short round wings can produce to get the bird off the ground in one flap. The primaries, spread out, are each clearly defined, and the impression is pushed perhaps two inches deep in the snow.

Here and there are the 'two together, two in line' tracks of rabbits of all sizes, and the even pawprints of a fox. In these conditions we may be lucky enough to see a fox, and if we are downwind of it, and keep still, it may come quite close. I remember being fascinated on one occasion by the sheer nerve of a fox in these circumstances. I was motionless, watching some Fieldfares

128

wolfing berries, when he appeared round a bend in the track: it took no imagination on my part to see that he appeared satisfied—content with the world and perhaps even humming a tune to himself as he came towards me. When eventually he noticed me, he was about thirty feet away, but instead of turning tail and bolting, he kept coming for a few paces before appearing to remember an important engagement in the other direction, turning quite cooly, and trotting back the way he had come. Only just before he had vanished from my fascinated sight did his nerve fail, and he cast a guilty look over his shoulder to see what I was doing.

Another nonchalant character lives overwinter in the hut, and sometimes we see him when the temperature outside drives us to eat our lunchtime sandwiches in shelter (along with the ever-present Robin, who shows no fear and needs no training when coming in through the door for crumbs, even though we know that it is a different bird almost every year). Against the background of our chomping jaws, mysterious scufflings begin in one corner behind the scythes and brishing hooks. Sooner or later, flickering whiskers appear, followed by a long narrow snout, and ultimately by a shrew—to judge from the length of its tail, a pigmy shrew. Shrews, as insect or other small animal eaters, must of course keep feeding right through the winter to keep their energy levels high enough to survive. Not for them the luxury of short periods of hibernation over periods of poor weather such as those the mice and voles can indulge in.

In amongst the rabbit tracks we can find the smudged passage of a squirrel—not surprisingly, with such short legs they try to keep in the trees as much as possible when the ground is snow-covered, but when they must descend, they try to get the nasty business over quickly, scurrying on splayed legs through a cloud of snow with tails held high and out of the way. In the last two winters we have had a completely albino squirrel in Northward Hill—and the all-white shape (quite large taking the fluffed tail into consideration) can give you quite a jolt as it ghosts quietly away along a branch as you approach on a late afternoon walk.

Perhaps the most extraordinary animal of all in Northward Hill appeared several winters ago. The warden at the time saw the beast first, and for fear of being accused of over-celebrating Christmas hardly liked to mention his sighting. He could think of no domestic cat with a regularly banded black and white tail—long and fluffy—and could only conclude that we had a visiting racoon. After several months of fleeting glimpses, we all became convinced of this—but the story has a tragic—and all too human—ending. The racoon was cornered on the marsh by a shepherd's two dogs—and he clubbed the unfortunate creature to death with his stick—for what reason we do not know, other than that the animal was strange. Such is the fate so often meted out to adders by irrational picnickers, and such was the fate of our last Great Auk, clubbed by superstitious fishermen on St Kilda in 1841.

Out on the marsh, the first handful-size skeins of Whitefronts are back—calling a lot when they are fresh-arrived, seeking the security of the largest flock. Until the Dutch marshes begin to freeze really hard, the bulk of our birds will not appear. The geese have accepted the quick change of almost half of the marsh from grazing to winter wheat: other birds remain, but have their difficulties. Now that the reeds and weeds edging the fleets are regularly chopped down, the Reed Buntings need to seek refuge far more in the wood, and usually repair there in the evenings, although most days they are still able to feed out in near-natural and more typical surroundings. Another surprise for the unwary on a December walk through the wood is the sight of Moorhens—quite a number of them, sometimes roosting in groups amongst the Pheasants. Many of these now seem to spend much of their lives within the confines of Northward Hill—how long they will be able to continue to do this remains to be seen.

Jays are at their most obvious at the moment—last month and this, any time you look out over the marsh, or farmland, you can see a Jay, slowly flopping away from the wood, looking unusually heavy-headed, or hastening back, sometimes calling harshly. The reason for the big-headedness is apparent once we get them in our binoculars: they are carrying acorns. These they will pick up from the leaf litter, or pluck from the branches before taking them out and burying them—apparently at random, in nearby fields. Apart from the scattering of food to be found at chance (it seems they have no 'intelligent' way of finding them again) by Jays or other birds or animals, this does ensure some spreading of oak trees when a few of the acorns germinate!

The ringing, of course, goes on. We can see by the birds that we catch that many of the youngsters are no longer about, and that accordingly the proportion of adults we catch is increasing. While it is rather misleading to calculate the 'average' length of life for birds, because of the high infant and juvenile mortality rate, it is interesting to see how long some of them are able to survive, so I have listed the 'oldest inhabitants' of Northward Hill (so far as we know them).

Those marked with an asterisk are migrants, with several thousand flying miles to face each year, as well as the normal hazards of a bird's existence. There are some unexplained surprises in this list: why do some of the larger birds (Great Spotted Woodpecker, Mistle Thrush) apparently have such a short span, for example, and why are Blackcap and Garden Warbler so different? Questions to be answered sometime in the future. Over 93 months (nearly eight years) you get to know a Blue Tit that you see perhaps twice each month!

Life-spans, in months, of some of the oldest Northward Hill birds

Little Owl	65	Blackbird	62
Green Woodpecker	49	Robin	60
Great Spotted Woodpecker	26	*Blackcap	83
Great Tit	79	*Garden Warbler	35
Blue Tit	93	*Whitethroat	23
Willow Tit	81	*Willow Warbler	35
Long-tailed Tit	38	Dunnock	70
Treecreeper	88	Bullfinch	78
Wren	38	Chaffinch	77
Mistle Thrush	29	Tree Sparrow	71
Song Thrush	60		

So far as the record is concerned, this is likely to be held by one of the Herons. So far as we know, none of our birds is still ringed (although they were just after the war) the national Heron record is of a bird living—in the wild of course—for twenty-five years. During December, we have a final task for the year in the heronry. When nests are being built in some years as late as August, leaf cover makes the count of the total number of nests difficult (although Dutch Elm Disease is rapidly solving this problem). So each year at about Christmas time, we have a final check and plot the positions of any new nests built since our last survey in late May or early June. Last year the colony reached 167 nests—making it by a long way the largest heronry in Britain and Ireland—and the steady increase from low levels after the 1962-63 winter continues, sometimes at as much as 10% per annum. Of course this heightens the importance of this colony—Herons are scarce in southeastern England, and the wood holds about 4% of the British population.

Walking around the wood it is obvious that there are fewer thrushes about— even if we stay on to watch the evening roost arrive. Partly this is because they have cleared up the berry crop and are more widely distributed over the farmland, partly it is because some mortality has occurred, but probably the major feature is that having paused and fattened in Northward Hill, many thrushes will have moved on. Sometimes this will be away to the west, but a film made by Marconi just across the river in Essex shows another possibility. Radar reveals bird movements very clearly—to such an extent that it causes problems to airport controllers in sorting planes from birds. The film, using time-lapse photography to speed the movements, covers a whole year in the Thames estuary, and shows clearly the early autumn bird arrivals from the east and northeast. From December to February there is an unexpected sight —echoes indicating many hundreds of thousands of birds, moving south-eastwards into Europe—and some of these may be our Fieldfares and Red-wings, together with Lapwings, all perhaps moving away from poor weather.

So the year in Northward Hill draws quietly to a close: over-sized Christmas lunches have been digested by brisk walks, disturbing thick-set Woodcocks from beneath snow-capped bracken, sending them twisting away on silent wings. Turkey soup and turkey sandwiches are the rule. Late in the afternoon the chill still persists in many places, especially on the paths, where winter has reduced to frost-fringed rosettes the bold thistles and teazles of summer. Only the Robin's winter song salutes the misty apricot sunset. What changes, what new insights, what surprises will next year bring? One thing is as certain as the sunset: the fascination will remain.

Postscript

Birdwatching can be equally enjoyable as a solo pursuit or with companions —this is part of its charm. While many of the observations in my diary have needed only my own presence in the wood, the more-detailed ecological studies have required a high degree of teamwork. To the various members of the team over the years I offer my thanks: David Musson, Jonathan Martin, John and Meg Bacon and, above all, Chris and Pam Cox. To the newest team member, Duncan Cox, this diary is dedicated.

Northward Hill is a National Nature Reserve, owned by the Royal Society for the Protection of Birds and at present managed, under agreement, by the Nature Conservancy. As in all Nature Reserves, access has to be rationally controlled, partly to allow the birds some peace, and partly so that (literally) the reserve shall not 'wear out'. Various factors may influence the access for visitors to the reserve, in part seasonal but, at the moment, part associated with dangers from the falling limbs of diseased elm trees, or from the felling operations under way. Up-to-date details of the availability of entry permits can be obtained from: R.S.P.B., The Lodge, Sandy, Bedfordshire. To avoid disappointment, it is absolutely imperative that permission is obtained *before* visiting the Reserve. At present permits cannot be issued on the spot by the warden.

WETLAND

by JEREMY SORENSEN

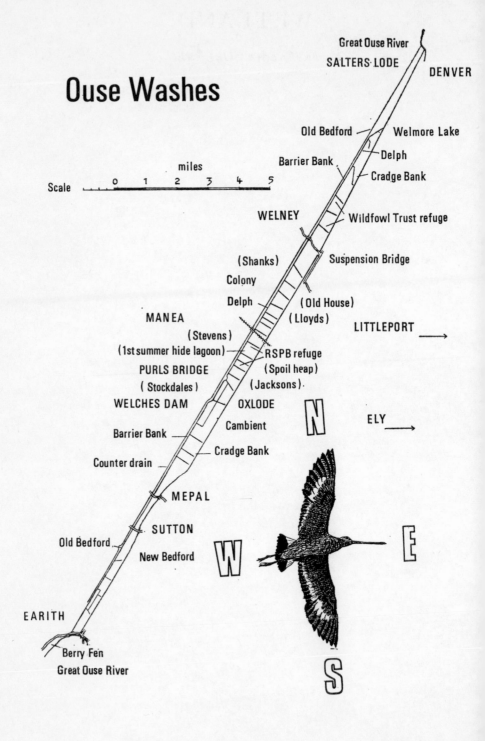

Ouse Washes

Scale

miles

0 1 2 3 4 5

Great Ouse River

SALTERS·LODE

DENVER

Old Bedford

Welmore Lake

Delph

Barrier Bank

Cradge Bank

WELNEY

Wildfowl Trust refuge

(Shanks)

Suspension Bridge

Colony

Delph

(Old House)

(Lloyds)

MANEA

LITTLEPORT →

(Stevens)

(1st summer hide lagoon)

RSPB refuge

PURLS BRIDGE

(Spoil heap)

(Stockdales)

(Jacksons).

WELCHES DAM

OXLODE

ELY →

Cambient

N

Barrier Bank

Cradge Bank

Counter drain

MEPAL

SUTTON

W

E

Old Bedford

New Bedford

EARITH

S

Berry Fen

Great Ouse River

Introduction

The Ouse Washes are about 20 miles long, and for much of their length half a mile wide. They stretch from Earith in Huntingdonshire through the Isle of Ely in Cambridgeshire to Denver in Norfolk, running NE/SW. It is a pity that Vermuyden when engineering the system in the 1650s did not think more about birdwatchers, because morning birdwatching can be quite difficult on the Manea side of the Washes when the sun shines into one's eyes. On the NW side the Washes are bounded by the Old Bedford River and on the SE by the New Bedford River. The rivers have high dyke banks running along their length. The agricultural land on either side of the Washes tends to be below sea level.

Washes, fields surrounded by ditches full of water, vary in size but most are between 10 and 30 acres. In some areas the individual washes are all quite large and run right across from the Delph river to Cradge Bank on the New Bedford River side. In other areas the ditch systems become quite complex and the individual washes tend to be smaller.

In the winter the Washes are a safety valve to the Great Ouse River system and in Summer are mainly grazed by cattle. Four conservation organisations own land on the Ouse Washes; the Bedfordshire and Huntingdonshire Naturalists' Trust Ltd, the Cambridgeshire and Isle of Ely Naturalists' Trust Ltd, the Wildfowl Trust and the Royal Society for the Protection of Birds. Between them they own over two-fifths of the land in the Ouse Washes.

Mentioned in the text are Russell Leavett an RSPB warden who assisted me on the Washes; Clifford Carson, another RSPB warden on the Washes; Brian Ribbands the Cambridgeshire and Isle of Ely Naturalists' Trust warden at the time; James Cadbury, senior RSPB biologist from headquarters; Josh Scott the Wildfowl Trust warden; Gareth Thomas another RSPB biologist from Sandy who is much involved with research work on the Ouse Washes; Ted Richardson in charge of the cattle in the area where most of the RSPB washes are; Ivan Palmer, a local bird ringer and wildfowler who gives us much valuable help on the Reserve.

I am indebted to Michael Allen for corrections and helpful criticism of the typescript.

Plants mentioned in the text are given the English names as found in Collins *Pocket Guide to Wild Flowers*, and birds the English names as found in BTO Guide 13: *A species list of British and Irish Birds*.

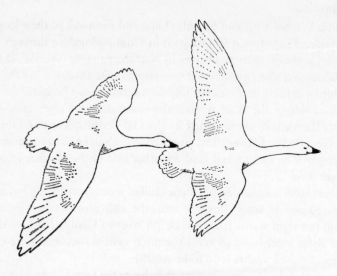

JANUARY

1st week. We are looking, Ivan and I, across the Washes over the top of Barrier Bank, by the bend at Purls Bridge. Immediately below the bank is the Osier bed alongside the Delph River—a mass of brown reed-like branches about six feet tall thrusting from the flood waters, the leaves long fallen. From the Osiers the width of the flooded river can be determined by the dead Dock and Reed-grass stalks just showing above the flood waters on the far side. The river bank still protrudes here and there as small islands above the slowly rising flood water, and odd groups of Wigeon graze the shrinking plots of grass. The flood waters stretch across the Washes about 200 yards back from the course of the Delph. On the less well-grazed washes dead vegetation, various grasses, show in confusion above the water. Where spoil from generations of ditch clearing has been deposited, double spits of land two yards wide at either side of a ditch reach like fingers into the flood waters for 50 yards or more, each pair about 150 yards from the next. These are covered with a multitude of wildfowl, the nearer ones with Bewick's Swans in line, preening and pseudo-sleeping with head tucked under one wing, some standing on only one leg. Close to the spits and in the shallows of the flooded washes hordes of duck, Coot and Moorhen swim about. Some are upending, some grazing near the edge of the flood just in or just out of the water, and others merely loafing about. The noise is impressive, with contended quiet honking from the Bewick's, the whistle of drake Wigeon, quacking Mallard, the bell-like quack from the Teal and the clamouring of Coot. Further back from the rising flood waters the Washes, with a slope so slight that it is unnoticeable, run 600 yards to Cradge Bank, which contains the swollen New Bedford River and prevents it from spilling into the Washes.

This New Year's day is dry and cool after a slight frost during the night. Although the flood water on the Washes is still shallow, the river level reads 156 cm (5 ft 1 in), just over four feet deeper than the summer level, and rises slowly by about four inches during the day. Conditions are ideal for viewing wildfowl between the railway and Purls Bridge, with the washes just under a quarter flooded, the deepest water averaging about two feet. Ivan and I walk along the Old Bedford River on Barrier Bank to estimate the duck and count the swans on the RSPB refuge area. The bulk of the swans are in a large flock on Jackson's and Stockdale's washes. Bewick's are quietly honking to each other in a contented manner and there is another small flock nearer the railway. The Bewick's total 363, 43 of them youngsters, an increase of 49 on the previous day's total (only two new young ones) fewer birds than on New Year's Day last year. Two Whooper Swans are also in the flock, both adults, and these are probably those seen three days ago, now back from feeding elsewhere on the Washes. The Mute Swans are the same as yesterday, twelve birds, seven of them immature.

We find good numbers of Wigeon on the refuge today, the larger flocks grazing where the washes have been more closely grazed by cattle during the summer; an estimated 7,000, well spread out at the back of the flood along the whole length of the refuge. Mallard total just over 2,000, with a fairly even wide-spread distribution. Groups of Pintail on the Spoil Heap and Steven's wash, and odd pairs sprinkled about the other washes, bringing the number on the refuge to about 50 birds.

At the end of the day we had only seen about 50 Teal, mainly in the newly flooded runnels at the back of the flooded washes, and the only Shoveler were five birds, four drakes and a duck, flying SW up the Washes towards Mepal. Two Pochard ducks were seen swimming in the deeper water of the Delph River, and nearer the railway we flushed a group of six Tufted Duck and two duck Goldeneye from the Old Bedford River; these flew over Barrier Bank into the Washes.

Whilst peering over the bank counting duck on the Spoil Heap wash we spotted a duck Smew standing very upright just out of water, an unusual attitude, and the bird had us fooled for a moment. As we watched, it took off and flew with a curious flight somewhat similar to a Coot's, with very quick wing-beats and a pied effect on the wings. We do not see many Smew on the Washes, which are normally too shallow for this species.

At the north east end of the refuge near the railway the four Gadwall I had seen the previous day were feeding in about the same place. Nearer the railway on the lowest wash of all a duck Goosander was swimming. During the walk along the Old Bedford River we flushed four Short-eared Owls from their roosts on the side of Barrier Bank, and on returning to Welches Dam there were two Little Owls in the pollarded Willows and a Kingfisher perched on the grille of the pump house.

For the next four days there was constant freezing fog with temperatures down to −6.5°C. All the flooded washland has frozen except where there are large flocks of swans. As a result, the wildfowl have concentrated in these unfrozen areas and the noise through the fog is an experience I shall never forget. The Bewick's Swans honk loudly most of the time (you can hear them clearly in Manea village almost two miles away) and in the fog it is possible to approach them without being seen. Together with all the duck noises it makes for a wonderful atmosphere.

The flood reached a maximum height of 169 cm (5 ft 7 in) on 4th and began dropping on 5th January. During the afternoon of 5th the fog lifted and the scene was fantastic, everything iced up or covered in hoar frost. The falling flood caused large areas of ice to sink with it, making loud cracking noises. The dead Docks and other vegetation looked like works of art in stainless steel and glass, the plants above water level thick in hoar frost, with a frill of ice at flood level, poised on stalks of glass. Throughout the 20 miles of the Ouse Washes the scene was the same. Except that we saw all our wildfowl again, the only birds of note this day were five Bar-tailed Godwits flying up the Washes towards Welches Dam.

Wednesday 6th, Brian, Gareth, James and I did one of our regular complete bird counts of the full Ouse Washes from Earith to Welmore Sluice. The water, still dropping, was only 119 cm (3 ft 11 in) deep. The washes still slightly flooded were almost completely frozen over; the only open water being in main ditches and rivers. A cold day, overcast and freezing hard.

Our complete bird counts are all made in the same way. James counts the stretch from Oxlode and Welches Dam to Earith, Gareth does the middle bit from Purls Bridge to Welney, and I count from Suspension Bridge to Welmore Lake and back along the other side to Welney Bridge. Brian, our fourth counter, works opposite Gareth to make sure that nothing is missed on the far side. We all carry large-scale maps of the Washes and record the number of each species counted and plot its exact position. We work behind the banks looking over the top so that the duck don't fly off, confusing the situation, and avoid counting at weekends when the Washes are disturbed by wildfowlers, and inexperienced birdwatchers who know no better than to walk on top of the banks. We count each wash individually and the results of the count are worked out from the maps later. The biggest area a counter tackles at one time is about

30 acres. We find we have no trouble from flying birds and, unless the Washes are disturbed by people, the birds are more or less in the same place on the return journey along the other side of the Washes.

The results of today's count are most interesting: 310 Mute Swans, 73 of them immature, a drop on last month; only seven Whooper Swans, all adult, but the Bewick's Swans are well up at 873 with 211 of them youngsters. Of the Bewick's, 393 were at Purls Bridge, 235 midway between Welney and Welmore Lake, and the rest evenly distributed between the other sections. The duck and wader numbers with the exception of the Wigeon are all down, a reflection of the very cold conditions and of shooting pressure, mainly in the two refuge areas. Because of the cold, Teal have gone down by over 2,000 since the last count, leaving only 240 birds. Only 68 Pintail and 60 of these were on the RSPB refuge. Pintail are a very dynamic species; immediately reflecting changes in weather or flood conditions by their movements. Very few Shoveler present, only 25 and all in the Purls Bridge area; this is a bird that needs open water in which to feed, and will soon move out in cold conditions. Over 22,000 Wigeon is most encouraging, the highest number ever recorded on the Ouse Washes at this time of the year, and the fact that 77% of them were in the refuge areas shows how important it is to have such sanctuaries for the wildfowl on major wetlands. Only a few diving duck are about as the water level is still too shallow for them: 108 Tufted Duck, 117 Pochard, 18 Goldeneye and six Goosander. As usual not many geese: 14 Greylag Geese and six White-fronted Geese. There are 1,400 Coot and 300 Moorhen. The wader populations have really dropped with eight Dunlin, 12 Ruff, only 113 Snipe, a Jack Snipe and a Woodcock.

2nd week. This week has been a little less hectic than the last. It began with milder conditions and a fifth of an inch of rain on 8th, the flood waters rising on 9th to 169 cm (5 ft 7 in) on our measure and the washes almost seven-eighths flooded. The Bewick's Swan flock at Purls Bridge passed the 600 mark on 8th but I was sure there must be over 1,000 on the Washes for the first time ever, so Ivan and I decided we must count all the swans on the Washes as soon as possible. We managed to do this on 11th and, sure enough, logged 1,110 Bewick's, 278 of them immature; they have had an excellent breeding season. The number of Mute Swans is still dropping, 254, with 74 immature. We also found a slight increase in Whoopers: 18 birds, four immature. A rough estimate showed a continued increase in the Wigeon at about 25,000, and Pintail have gone up again with over 1,000 birds present. We stopped at Welney on our way back to let Josh Scott know the good news.

A few waders have been passing through, with two Golden Plover and three Grey Plover on 8th and 9th respectively. I also heard both species fly over on the night of 10th—a superb call the Grey Plover's, one of my favourite flight calls. We also had a Curlew, two Redshank and 20 Ruff on 10th.

On 14th Gareth and I arranged to do a dawn to dusk watch on the wildfowl and to record their feeding methods, timed at half hour spells throughout the day. We punted down the river an hour before dawn so that we could get into one of the forward hides without disturbing the duck. We managed this successfully and settled down to the chore, but two hours after dawn we had to give it up because of thick mist! Nevertheless, we had a bit of excitement that night, when a wildfowler came off the Washes with a Falcated Teal in his bag, doubtless an escape.

3rd week. This week the Bewick's flock has been slowly dropping at Purls Bridge with 517 birds on 16th. A female Hen Harrier over the RSPB refuge on 15th and 16th was a superb spectacle, causing areas of panic amongst the multitude of wildfowl. On 18th Gareth, Ivan, James and I had the international wildfowl count to do. In the usual way, we mapped the distribution of wildfowl and waders and recorded all other species. It was a mild day with a little sun and not too much wind; the floods were dropping, quite low at 124 cm (4 ft), ideal for a count. The Bewick's beat all records again with 223 up at Earith, 127 at Welches Dam, 423 at Purls Bridge, 188 below the railway line and 273 below Welney; a total of 1,234 Bewick's Swans, 276 of them immature —quite fantastic. The Mute Swans continue to drop with 259 birds, but the number of immatures has risen to 80, so obviously there is a population change of some kind going on. A reasonable number of Whoopers present; 19 adults and six youngsters.

The duck have increased significantly since the cold spell, the refuges containing 80% of the total duck population. However, Teal are spread evenly all over the Washes and we have never seen so many, 4,660 recorded. Pintail numbers are quite good at 1,500; this species seems to favour certain washes on which one finds large flocks with thin distribution elsewhere. Down below the Old House one wash had 122 Pintail on it. Shoveler numbers are looking a bit more respectable with 166 counted; Gadwall good at 61 birds; Wigeon are as spectacular as Bewick's with 32,000 birds present, over 17,000 of them packed into the RSPB refuge, quite a sight. Still very few diving duck, the floods are far too shallow, only 12 Tufted Duck, 63 Pochard, four Goldeneye and five Goosander. The only goose of the day a Pink-footed Goose at Purls Bridge.

The wader numbers are small but becoming more interesting. We counted 45 Golden Plover on the high ground near the Welney Road; these should build up well during the next few weeks. There are 480 Lapwing, a good sign of thawed-out land, four Dunlin and 29 Ruff, 20 of the last round a pool left behind by the flood near the Mepal road. The Redshank seem to disappear in the winter, only four seen all day. Gareth had five Curlew near the Old House, but between us we only managed eight Snipe, so they have yet to come back. James had a Jack Snipe up near Mepal. Only two Short-eared Owls and four Kestrels between us, so their numbers are well down since the autumn.

We counted 480 Fieldfare between us, many feeding on Cradge Bank, but only 11 Redwing. At different times I had a Twite and a Redpoll fly over whilst counting my section. James had a flock of 50 Greenfinches in a rough wash near Mepal and a flock of over 1,000 Tree Sparrows in some even rougher washes near Sutton. Altogether a very satisfactory day with 66 species recorded between us.

4th week. On 22nd the water started to rise again, reaching 170 cm (5ft 7 in) by mid-day, so there must have been quite heavy rain last Wednesday/ Thursday in the Midlands. On the night of 23/24th we had almost an inch of rain, and with a full moon on Sunday 30th indicating a spell of high tides, it was obvious that there was going to be quite a flood. The New Bedford River cannot cope with all the water so the sluice gates at Earith have to be opened and the excess water flows into the Old Bedford River, thereby flooding the Washes.

Still 300 Bewick's at Purls Bridge on 23rd, but from experience we knew they would not stay once the water went above the 200 cm (6 ft 7 in) mark; a very full flood. Sure enough, on 25th they moved down to the Cambient wash in the refuge.

On 27th we did another count of the Washes to get the wildfowl distribution plotted at the high flood level and before the end of the wildfowling season. Quite a day, with a strong SW wind and rain all the time we were counting; it cleared up as soon as we finished! The water was up to 284 cm (9 ft 4 in) and still rising, the roads at Earith, Sutton Gault and Welney were all covered with the flood waters. Brian, Gareth, James and I were out counting, doing our best to keep our optics dry, and especially the paper, otherwise our work might disintegrate before our eyes!

The Bewick's Swan numbers had dropped to 889 of which 188 were youngsters. Of the total, 387 were packed on to high ground at Earith, 308 were sticking it out on the Cambient washes at Welches Dam, and 96 round the Welney road where it was easy to get close in a vehicle. The other 127 were spread in small flocks and about to be moved on by rising floods. There were 23 Whooper Swans (only one immature), 16 of them on the very rough washes near the anti-tank ditch at Oxlode where there is a lot of emergent vegetation. All but 80 of the 308 Mute Swans were up at Earith. The dabbling duck population was already showing the effects of the high flood, reducing in number to only 3,000 Mallard, 300 Teal, 39 Shoveler and a Gadwall. The Wigeon population, able to graze the high banks, is more resistant to moving off at times of high flood, and 28,000 remained. Diving duck had yet to find the flood and numbers were still low. We had a flock of 23 White-fronted Geese at Purls Bridge from 26th to 29th, but they had gone somewhere else during our count and were not recorded on 27th.

The first Black-tailed Godwit returned on 28th and spent its time feeding and resting on the higher washes near the Welney road. The rising waters have been noticed by the Little Grebe and we counted 10 amongst emergent vegetation at Welches Dam on the Saturday night when the water was up to 320 cm (10 ft 6 in). Strong winds at the end of the month blew in three oiled Red-throated Divers on 31st and probably brought in the 15 Shoveler on the same day, though we regard it as normal for this species to come in about now.

The last day of the shooting season and many wildfowlers were unable to retrieve their quarry owing to the high flood; we found 25 Wigeon, eight Teal, six Mallard, three Pintail, two Pochard and 11 Coot washed up on the tide line. The flood remained steady on 31st and seems to have stopped rising at 325 cm (10 ft 8 in).

FEBRUARY

1st week. The flood has been dropping all week, falling from 300 to 239 cm (9 ft 10 in to 7 ft 10 in), but with the flood still so high that there are few wildfowl to see at close range from Purls Bridge. The area looks like a large new reservoir, with just the tips of the Osier branches showing through the flood and choppy water lapping halfway up the Barrier Bank. However, the high ground near the Welney road is superb territory during the week with hordes of wildfowl, especially Wigeon, and about 50 Bewick's quite near the road. These are pushed further off at weekends by eager, inexperienced birdwatchers getting out of their vehicles to 'get a better view', or walking on top of the high banks, skylining their silhouette. A pity, because neither they nor people following enjoy the scene as it should be.

The two worst-oiled Red-throated Divers must have died, the third does not seem too bad. We had good views of this one on 4th and 5th. The other two had been preening most of the time but this one spends its day mainly fishing or loafing around. The Pochard at last seem to have noticed our flood: a raft of about 600 near the shelter of the railway embankment, together with about 1,000 Coot. Some of our visitors express a difficulty in identifying Pochard from Wigeon at long range. I find the quickest way is to note the silhouette of the tails when the ducks are sitting on water; the Pochard's goes down into the water and the Wigeon's comes up out of it.

The Wigeon are providing an excellent spectacle in the evenings, large flocks of many hundreds flying up and down the length of the flooded washes at various heights, some very high indeed, in constantly changing V formations with much whistling from the drakes, the growl of the ducks and occasional wing clapping when birds get too close.

Some small mixed flocks of finches feeding at the 'tidal edge' along Barrier Bank, mainly Goldfinch and Linnet with one or two Twite amongst them. Meadow Pipits and Reed Bunting feeding on flotsam caught up in the flooded Willows: every now and again they break through the floating surface and have to flit quickly to safety.

A Great Crested Grebe came into the pit at Purls Bridge on 2nd; it will not be long before they are displaying. There is more activity on the Heronry than ever before. With luck we will have an increase on last year's two pairs. Not worth risking a close look and putting them off at this stage in the proceedings.

The gull roost seems to be getting bigger; I have yet to figure out a way of getting a reasonably accurate count of the gulls roosting on the whole Ouse Washes. At the moment, coming in at Purls Bridge, there are at least 25,000 Black-headed Gulls, a good 200 Herring Gulls, over 500 Common Gulls and at least 40 Great Black-backed Gulls. With roosts at Welches Dam, Old House and below Welney, counting would be quite a task at any time, but they come in so late there is not time to get round all the roosts before dark.

2nd week. Days of damp mistiness at the start of the week, but at least there was no proper rainfall and the flood continued to come down. On 11th the water was down to 196 cm (6 ft 6 in) and sure enough the first Bewick's started to come back to Purls Bridge—40 arriving on 11th, up to 200 the next day and 350 by 13th. It seems that 200 cm (6 ft 7 in) is the critical depth for Bewick's at Purls Bridge. If the water is deeper than this they move away to higher washes. One of the most distinctive identification characteristics of the Bewick's Swan is the way it lands, which is quite different to the manner of the two larger swans. It comes in from a much greater height, circling round and dropping in on down-curved wings, with the neck at an angle forward and slightly upwards from the body, a slight bend in the neck, as though parachuting down. The two other species come into land more like heavy aircraft; low, straight and steady. If there are already Bewick's there they often go through a sort of greeting display, or possibly a display to establish pecking order. It all depends on whether they are members of the family or old rivals. Each bird stands (or floats) with head and neck outstretched, wings open and slowly flapping—all the birds taking part grouping round with heads to the centre honking, trumpet-like, pumping head and neck up and down.

During the last few days it has been possible to get some excellent views of Water Rails near the Welney road; looking SW up the Delph river from the Welney bridge they can be seen foraging out on the open washes to the left of the Osier beds. Water Rails rarely stay out in the open so long, but of late there have often been two or three and we have seen up to seven on occasion.

Fortunately the weather became quite respectable at the end of the week, in time for our next full count on 14th. Excellent visibility with a light cold wind, a shade south of east; ideal. James, Gareth, Brian and I were counting. Another excellent result with Bewick's Swans again beating all records at 1,278 (210 of them immature); 28 Whooper Swans, and 334 Mute Swans. The duck were interesting, needless to say; Mallard almost 6,500 counted, 2,960 Teal, 1,400 Pintail, Shoveler coming up well as they do in the late winter at 1,050, 37 Gadwall and 24,500 Wigeon.

At last the diving duck numbers are respectable, 460 Tufted Duck and 5,480 Pochard, with over 5,000 in one raft just above Welches Dam; but only 19 Goldeneye—a direct result of the recent high floods and water still right across most of the washes at a depth of 183 cm (6 ft). Coot are well up with 3,870 birds counted. Nine Great Crested Grebe, 20 Herons, 10 Short-eared Owls and 13 Kestrels deserve note.

The waders are still a bit low, but showing signs of a pre-spring build up: 156 Golden Plover, 1,240 Lapwing, 32 Redshank, 36 Ruff, 102 Snipe, one of which was drumming, and 28 Dunlin. The distribution reflects both the recent high flood and the end of the wildfowling season—most of the duck being on the shooting zones.

James had a Dark-bellied Brent Goose which must have got a bit off course, and between us we recorded 66 species including 33 Carrion Crows, one Siskin, 64 Goldfinches, two Greenfinches, five Chaffinches, two Corn Buntings, 47 Reed Buntings and four Yellowhammers.

3rd week. The new moon at the beginning of the week caused a series of high tides which must have held up the flow in the New Bedford River to such a degree that the sluice gates at Earith opened, releasing more water into the Old Bedford River, because the flood level crept back to 200 cm (6 ft 7 in) by 17th and only dropped back to 195 cm (6 ft 5 in) by the end of the week. The swan flock at Purls Bridge managed to hold on but the number of Bewick's was reduced to 420 by 17th; however we had 20 Whoopers on that day and 60 Mutes making it an excellent place to learn swan identification.

On 18th Clifford, driving to my house, had to brake hard to avoid a Bittern walking along the drive! which is the second time I've missed seeing a Bittern in my garden. Last year a local farmer saw one in the ditch on New Year's day, but we get very few as there are hardly any reeds in the area.

We are seeing a lot, in the evenings, of a pair of Barn Owls that fly up and down the road between Welches Dam and Purls Bridge. Another bird we seem to do well for is the Stonechat, a pair spending the winter on Barrier Bank between the railway and Purls Bridge. There is a pair of Stoats breeding under an old plank down there too. On this particular habitat Stoats pay their way. The winter floods control their numbers so that the population of Stoats during the bird breeding season does not cause excessive damage, yet we have the advantage of the Stoats' controlling influence on the Washes rat population. I doubt if we could cope satisfactorily with that ourselves.

It is interesting what one learns wandering about on the Washes after dark. In a fairly high hay wash we found Skylarks, Reed Buntings and Corn Buntings all roosting in the open, well spread out in smallish groups, mainly in the thicker stands of Reed-grass which occur in this particular wash.

The Black-tailed Godwits seem to be coming in early this year, a flock of five on 17th and up to 19 by 21st, all messing about down near the Welney road.

4th week. This is hardly winter weather—dull and mild all week, the wind staying in the NE the whole time and never very strong. One consolation has been the small amount of rainfall. The flood has been going down all week from 192 cm (6 ft 4 in) to a depth of 146 cm (4 ft 10 in) by the end of the month.

On 25th, James, Gareth, Clifford and I made our month-end count of the Ouse Washes. Another day of good visibility with light NE winds and a flood depth of 172 cm (5 ft 8 in). The results of the count illustrate two things well when compared with last year: first, the stability of our Bewick's Swan and Wigeon populations on the Ouse Washes as a whole, with the local variations showing that no one area of the Washes is especially important; second, that the whole Ouse Washes system is used by all species of wildfowl at some time.

Our count showed that all species of swan had increased in number, with 361 Mute, 35 Whooper and 1,280 Bewick's—probably the pre-migration peak we have at about this time each year. The Mallard were low at 2,330, but we noticed many feeding in the fields of winter wheat on the fens surrounding the Washes. Actually, when studied closely, these duck were not taking the wheat but were eating the old potatoes left in the field from the previous harvest and rotted by the frosts (potatoes being the previous crop in the rotation of the fields in question). Pintail had gone up to 2,050 and, in addition there were good numbers in with the Mallard on the fen fields, so this count recorded less than the actual number of these two species using the Washes. Shoveler were surprisingly down to 470, Teal to 940 but Gadwall showed a slight increase to 46. Wigeon were stable with 25,900 counted. Diving duck were holding on well—280 Tufted Duck, 4,570 Pochard, a Scaup, nine Goldeneye, three Red-breasted Merganser and four Goosander.

There was a selection of geese at the Welney end, 22 Greylag, 34 White-fronted and two Bean Geese, adding that bit of spice to things. Waders were coming up well, pre-spring: 400 Golden Plover, most near the Welney road on their favourite washes, 2,100 Lapwing, 57 Dunlin, 15 Redshank, 87 Ruff (most of these on some flooded ploughed washes near Welmore Lake), and 12 Black-tailed Godwits near the Welney road, the only ones we could find. Only 40 Snipes, seven of them 'drumming'—the territorial flight display, during which the bird flies about as though riding a 'big-dipper', opening its outer tail feathers when it dives. The wind passing over these feathers produces a noise similar to the bleating of a Goat.

On the last day of the month a flock of 20 Bramblings came in to feed on the floodline at Welches Dam. Their richly coloured mantles and white rumps giving an impression of blown autumn leaves as they flitted along the water's edge.

MARCH

1st week. Clifford and I spent the first three days of the month cutting and setting Osiers down by the first summer hide. The washes are still flooded but the water level is dropping. We got rather wet on the Wednesday as the water was a bit high for working in, at 142 cm (4 ft 8 in), but it will soon be the breeding season and we must have the job finished. We set over 1,000 new Osiers from those we trimmed back. I like the way a six-foot branch of Willow can be cut and trimmed, sliced into three two-foot pieces and each one set. They almost always grow, too. It was even worth a soaking because the ground, being well softened by the water, accepted the new Osiers easily.

On 4th we had over 60 Black-tailed Godwits spread out in three main flocks between the railway and the Cambient washes at Welches Dam. This is very promising as we only had 14 this time last year, and the year before that they did not arrive until 10th of March. The weather is very mild and this may have something to do with it. In previous years they have usually flocked nearer Welney at this stage, so this season more may breed up Purls Bridge way.

The sunrise on the morning of 6th was out of this world. The washes still slightly flooded, the sky rich orange with flame yellow clouds streaked across it. Osiers stood in silhouette in front of the sky-coloured Delph river, and beyond it most of the far bank, just out of water, lay as a band of dark colour between the river and flame-coloured floodlands on which 200 Bewick's

Swans tinted by the sky were awakening from roost. Above them a constant stream of hundreds of gulls, lit by the mass of colour, flying up the Washes from roost. The far horizon a dark relief, with individual leafless trees depicted at great distance with a light, hazy look. Needless to say the day ended very wet, the rain gauge recording 8.8 mm, about a third of an inch. This pushed up the water level by the end of the week to one centimetre above that at the beginning.

The Great Crested Grebe have been displaying regularly in the pit at Purls Bridge and on the Delph river opposite. There have been some excellent views of the most spectacular display—a pair will dive together and each come up with weed then swim towards the other, suddenly rising, to stand almost on the water breast to breast for a few seconds swaying from side to side. Much head shaking display throughout the day too.

2nd week. The weather has done almost everything this week. Starting off dull and mild, a real wintry day on 11th with snow squalls on a strong east wind, the next day cold, the wind gale force but otherwise fine with some soil blowing off the fen fields, and the week finishing fine and mild, the wind southerly and the temperature reaching 15°C. The water level rising fast to 210 cm (6 ft 8 in) on 11th and dropping to 190 cm (6 ft 3 in) by 14th.

Much of our time has been spent preparing for the Summer visiting season, putting gravel on car park areas and fencing round hides before the cattle begin grazing, and getting the boat ready for launching.

We were amazed to notice that one of the Herons has two large youngsters already. They must have laid eggs in the middle of January! The Bewick's Swans are starting to leave, small flocks flying mainly NE down the Washes, but 999 (!) were still with us on 8th. The birds actually leaving can be distinguished by the immaculate V formation; the normal day to day flocks flying in a more chaotic undisciplined way. Various signs of spring are evident: the ducks all in pairs with numerous groups of up to a dozen drakes chasing a lone duck. Mallard, Gadwall, Wigeon, Pintail, Shoveler and Teal all doing it. An unpaired duck is easily identified from a paired duck being chased. In the latter case the drake of the pair can be picked out in the chase because it constantly manoeuvres to ensure that it places itself between its duck and the other drakes in the chase, very successfully most of the time, too. The display flight is interesting, the drakes always making much noise, the flock weaving all over the place.

Some of the waders are also showing signs of spring. Quite a lot of Snipe drumming at the end of the week. Redshank in parachute display song, Lapwing weaving about the sky and Black-tailed Godwit doing ceremonial song flight displays. The Washes are an unbeatable place on still warm days at this time of the year. A group of Ruff makes a half-hearted display from time to time, some birds showing signs of summer plumage, well ahead of all the others. Small flocks of up to half a dozen Twite and a dozen Brambling have been passing over during the week. A pair of Barn Owls quarters the Old Bedford low bank early morning and late evening; Little Owls call in the copses at Purls Bridge and Welches Dam, and two Short-eared Owls are present, still.

3rd week. What a fantastic spell, fine clear days, frost at night and misty mornings. The temperature reached 21°C on 15th and 20th, the flood dropping all week from 187 to 159 cm (6 ft 2 in to 5 ft 2 in).

On 17th James, Clifford, Gareth and I carried out another full count of the Washes. A late start, the mist not clearing until after ten o'clock. We found only one Whooper, but 251 Mute and 670 Bewick's Swans. Duck numbers holding well: 16 Shelduck, 2,370 Mallard, six Garganey, 100 Gadwall, 1,620 Teal, 2,050 Pintail, 1,080 Shoveler and 21,200 Wigeon. The flood still holds a respectable number of diving duck: 400 Tufted Duck, 870 Pochard, 16 Goldeneye, a Scaup, 1,530 Coot and 510 Moorhen. There was a Red-throated Diver in good condition at Purls Bridge. Two Water Rail, 28 Great Crested Grebe and seven Little Grebe.

An interesting day for waders: 1,570 Golden Plover, four Ringed Plover, 1,060 Lapwing, 112 Dunlin, 225 Redshank, 112 Ruff and 115 Black-tailed Godwits most on the Cambient main washes. Things are looking great for the breeding season: 790 Snipe counted, 96 of them drumming, and two Oyster-catchers near where they bred last year.

Some of our other results included 1,240 Woodpigeon, four Kingfisher, 188 Skylark, 75 of them singing and 140 Meadow Pipits of which 48 were singing: 390 Fieldfare, most feeding on Cradge Bank, and 305 Redwing—one so tame it almost gets under one's feet before flying on a few yards. Altogether 77 species recorded during the day.

The late start caused a late finish and we were counting the Swans from the Wildfowl Trust hide as Josh came to feed them. A fascinating sight, with most of the swans loafing around the side of the pool preening or roosting, and a few swimming about in the centre. As feeding time approached the honking noise and activity increased steadily. When Josh came out from the bottom of the hide with the barrow load of cereals the mixed herd came swimming across as he wheeled the barrow along the water's edge in front of the hide. He spread the cereals on the water and bank behind him, and the Bewick's were coming as close as the Mute Swans, to within touching distance, in their eagerness to feed. We counted 338 Bewick's by the hide together with 98 Mutes, not forgetting a Black Swan also present.

Next day brought the arrival of our first passerine summer migrant, a

Chiffchaff, singing in the Osier bed up near Welches Dam at dawn. The last three days has seen much activity and chasing among the Black-tailed Godwits and the ceremonial display song can be heard much of the time. This is a most moving performance in which the male flies with staccato wing beat to a height of 100 to 200 feet, uttering slow disjointed 'wi-ka wi-ka wi-ka' call. On gaining the required height, flight and song change. The flight is extraordinary; the bird appears to be flapping its wing alternately (in fact it is together) in a jerky rhythm, the tail twisting the whole time from side to side, causing the body to swing from side to side. The call is now a superb lonely sounding 'whut-to whut-to' noise carrying for great distances across the marsh. This song flight, sometimes going on for many minutes, is completed when the bird suddenly closes its wings and dives fast to the ground, opening its wings at the last moment to land, wings upheld for a few seconds, contrasting white against the green grass. The wings close and the bird merges into the background.

4th week. The floodwaters dropped from 147 cm (4 ft 10 in) on 22nd to 58 cm (1 ft 11 in) at the end of the month. On 26th it finally came off the washes to a depth of 97 cm (3 ft 2 in), leaving the fields covered with small splashes of water in the runnels, the grips in the centre of each wash and the ditches at the side pouring surplus water off into the Delph. The fine spell continued until 27th when we had a day of squally westerlies gusting to gale force, causing some of the fen fields to 'blow' a little. The month ended with rather dull mildish conditions.

By the end of the week it was obvious that some Snipe, Redshank, Lapwing and Black-tailed Godwit had eggs, though we knew the exact location of none of the nests, it being a little too cold to look anyhow. Mallard and Shoveler now well advanced with the first Mallard chicks expected any day now.

No sign of any more passerine migrants until 25th when we had our second Chiffchaff passing through. An interesting flock of 23 Great Crested Grebes at Purls Bridge on that day too; pre-breeding dispersal I expect. They must be more northerly breeders, though, because the pair at Purls Bridge is incubating eggs. A flock of 100 Twites with equal numbers of Linnets, plus a few Greenfinches, on a crop of radishes by the Counter drain at Welches Dam is interesting from our point of view, though I doubt if the farmer is pleased.

Another full count of the Washes on 29th, the usual four of us counting. The floods have receded, leaving a few very shallow pools and damp areas, mainly on the Delph river side of the washes.

The Swan numbers are dramatically down with 305 Mute Swans, 130 of these in a non-breeding flock at Earith. Only 26 Bewick's, five of them immature. Duck numbers dropping with breeding dispersal well under way: 25

Shelduck, 1,970 Mallard, four Garganey, 36 Gadwall, 860 Pintail, 2,200 Teal, 710 Shoveler, 5,020 Wigeon, 200 Tufted Duck and only four Pochard. Still 1,250 Coot and 530 Moorhen.

With the exception of passage species our wader numbers are mainly the breeding birds: 710 Golden Plover, 478 Lapwing, all seem to be paired and on territory; 150 Dunlin, 237 Redshank, 100 Ruff and 168 Black-tailed Godwits. We counted 1,000 Snipe, 388 of them drumming. A Jack Snipe and a Curlew rounded off the waders. James had a Lesser-spotted Woodpecker drumming in a copse near Mepal. Still 730 Fieldfare present and 139 Redwing. Sixty-seven species in all recorded between us.

APRIL

1st week. April fool's day. A Mallard with a brood of five chicks on the first hide lagoon, the first we've seen this year. Not the best of weather for rearing chicks, strong winds and still quite cold. A busy weekend for us with the Easter crowds to worry about. A delicate part of the breeding season, this, with the waders and duck beginning their clutches. There are always one or two birdwatchers who just will not realise the consequences of their actions and we have to be on the alert or we would have birds deserting nests. Fortunately, most birdwatchers are keen to cooperate.

Already some Lapwing, Redshank, Snipe and Black-tailed Godwit are incubating full clutches of eggs and we found our first Shoveler nest on 2nd, seven eggs in a tussock of Great Pond Sedge. Very little cover for the nests at this stage, as the grass grows very slowly and patches of the washes are still very wet. We also found a new breeding species on the reserve on the same day, a Tawny Owl with four eggs, nesting in a beer barrel we had erected in a tree (for Barn Owls) some two years ago. Both adults were in the nest and among the prey in the nest was an almost complete Water Rail; interesting. Easter Monday was a great day, two Heron chicks fledging from the Heronry, already! It seems there will be six nests this year, an increase of four on last year.

In the evening we were in the first hide, and at the back of the lagoon abo
50 yards away, on an area of damp mud, was a flock of 98 Ruff feeding. Qu
often sections of the flock would break into display, usually when one of t
nine Reeve feeding with them came near. The Ruff are still in fairly po
plumage with the exception of one or two individuals. An all white bird whi
(assuming that it is the same bird) seems to come into summer plumage ear
each year. And a gorgeous specimen, with a copper coloured cap running
a point down it's neck, and a white 'Elizabethan collar' and breast, the man
mottled ginger, orange legs. A third bird with bright ginger head, black coll
and breast, and almost red legs. These seem to be the dominant characte
The tilting displays are somewhat reminiscent of knights of the round tab
other birds stand about quite still for many seconds with beak pointing eart
wards all feathers erect.

Further to the left from these Ruff a group of seven drake Shoveler a
making their 'gudunking' noise with heads bobbing. One drake keeps betwe
the other six drakes and the lone duck with them; eventually, all fly off wi
the one drake keeping station with the duck the whole time as they weave
position in the sky; all the drakes still 'gudunking' and eventually the otl
six drakes fly back, one by one, to the lagoon.

Between these Shoveler and the Ruff a couple of Dunlin, a Spotted Re
shank in winter plumage and a Ringed Plover are feeding quietly in t
shallow water. As it gets near to dusk a couple of Black-tailed Godwits fly
and start to feed, wash and preen busily.

The 7th saw a number of summer migrants coming in with four Swallo
a Willow Warbler singing in the Osiers, and six Yellow Wagtails feeding
Barrier Bank. Still two Fieldfares about but no Redwings, perhaps the sev
of the previous day will be the last we shall see until the autumn. The weste
gale we had on 6th must have washed out the Great Crested Grebe's nest
the pits at Purls Bridge because both birds were swimming about together
day on 7th.

2nd week. Dull, and a bit cold for the time of the year, sums up the weatl
for the week. At least very little rainfall so the river should stay put. We l
to build a screen of dead Osiers by the first hide so that our visitors could
it without scaring all the birds on the lagoon. The natural Osier screen
planted is not too efficient at the beginning of the season until the leaves co
out more fully.

No more summer migrants came in this week. The Ruffs are very act
with between 30 and 50 busy on the RSPB refuge area, but only four Reeve
most, which is a bit worrying. The Great Crested Grebe started to re-lay
9th in the same place as before. The Black-tailed Godwit situation is m

encouraging: it looks as though over 30 pairs will breed between the railway and Purls Bridge, so if we don't have a good year it will be because other parts of the Washes have fared very badly. A flock of 112 Icelandic Black-tailed Godwits came through on 10th and 11th. These were in a flock among the Godwits nesting on the Cambient washes at Welches Dam. The Icelandic birds are smaller than ours, with brick red on head and neck, whereas ours are more of a cinnamon colour; it is easy to see the difference when they can be compared together.

A Black-tailed Godwit which nested near the first hide lost its eggs to a Carrion Crow on 12th. During the early part of the season the Crows do well whilst the grass is still short.

James, Clifford, Gareth and I did a full count on 14th. Poor visibility beyond 400 yards, the river depth 68 cm (2 ft 3 in), normal highish. There are 272 Mute Swans still with us, almost 200 of these as a non-breeding flock at Earith. The only wild swans were two sick Bewick's at Earith and two at Welney.

The duck counting was complex as we were recording drakes and pairs rather than just counting: 26 Shelduck (14 pairs), 970 Mallard, 70% of these drakes, most ducks either incubating or with small ducklings; 240 Teal (only 15 pairs), seven pairs of Garganey, 73 Pintail (42 drakes), 473 Shoveler of which over 60% were drakes, most of the ducks we saw probably had started clutches by now; 46 Gadwall, half drakes, late breeders. Still 2,950 Wigeon but I have yet to come across a breeding duck of this species here. All reports followed up so far have proved groundless. The Tufted Duck mostly in the Delph, 158 counted. Only one Pochard, 310 Coot (last year we had 700 at this time) and 250 Moorhen counted.

The waders are becoming very difficult to count now. Many are incubating eggs, some just laying and others yet to set up territory. We counted 450 Lapwing (202 pairs); the Redshank look good, 221 (100 pairs), Black-tailed Godwits excellent, 33 pairs located already between Mepal and Welney; 112 Godwits seen in all; 497 Snipe, and 447 of these were drumming; one pair of Oystercatchers. Other waders on the count include two Ringed Plover, 370 Golden Plover, 54 Dunlin; only 14 Ruff could be found and all these at Purls Bridge, two Greenshank and a Spotted Redshank. In all, 69 species recorded on the count.

3rd week. Conditions are still rather cold for the time of year, with frost most nights this week, down to −5°C on 17th, which won't do the Mallard ducklings much good. This weather certainly slows migrant passage. The only new birds during the week a Sedge Warbler on 17th singing away in its familiar spiel, sounding very irritated about something, and a Common Tern the next day. The duck numbers are holding well on the RSPB refuge area, 100 or so Teal, about 1,000 Wigeon and 40 or so Pintail. There is no doubt that the water retained on our summer lagoons makes a significant difference to the breeding duck and waders.

Quite a lot of lekking display by the Ruff in the very early mornings, and their plumage is beginning to get really interesting. Between 20 and 40 Ruff taking part, but we have never seen more than five Reeve. The Ruff seems not to like lekking in the high winds, but they certainly favour the wet areas we are keeping for them.

I have never seen so much activity from the Black-tailed Godwits between Purls Bridge and the railway, at least a pair on every wash, and some washes have as many as four pairs; to think that four years ago it was a job to find more than six pairs on this section. As with the Ruff, it is the wet areas we have made which pull the birds into this area.

A good 600 Golden Plover on the drier parts of the washes between Purls Bridge and the railway; normally we find them nearer the Welney road so its a change to have them up here. A Greenshank on the lagoons on 19th and 50 Dunlin commuting between the three wet areas in the refuge. A pair of Pochard in the Delph opposite Welches Dam; I hope these breed as there has been at least one pair each year since I came here. A Corn Bunting is singing in the field near the house at Welches Dam, and both Little Owl and Barn Owl are present in the small copses round about. The Barn Owls fly past the hides on the Barrier Bank regularly in the evenings. There seem to be more Common Partridge about this year; I suppose Red-legged Partridge must be about 25 to 1 over the Common Partridge, but two pairs of Common Partridge are spending more time near the house and we are seeing much about generally. Last year on 15th I remember that whilst shaving in the morning I happened to glance out of the window of the caravan I lived in then, and was amazed to see a Firecrest feeding in the privet hedge no more than six feet away.

4th week. Still it continues cold, the grass is behind this year and as a result the cattle are late coming on to the washes. The first herd arrived in three large floats on 22nd: 92 head of black bullocks imported from Ireland, to graze on washes near the Old House wood. Still, at least its very dry and the river is low. This probably explains why there are many less Coot nesting than in previous years when we have had short spring floods. Last year's began on 27th, all the Coot frantically building up their nests to keep them above the rising flood waters, then (a thing I find so amusing) unbuilding them as the flood receded, with the resultant huge pile of material next to each nest when the flood had gone, making each nest site obvious. Some Coot are hatching as we see April off; not a successful species on the Washes, though. They sort out territories when the place is still flooded (a suitable habitat for the bird), nest on a receding flood, have young when the Washes are dry, and predate each others young. They are able to lurk behind tussocks unseen by the parent bird until it's too late, whereas the chicks would be safe from this on open water.

One Mute Swan, obviously a pessimistic old-timer, nesting on Shank's wash, has built a massive nest this year. It will take quite a flood to see it off, complete with its six eggs.

This has been a reasonable week for wader passage, with Ringed Plover passing through from 24th, seven on 28th, and three Little Ringed Plover on 26th, all on the lagoons. Between 300 and 500 Golden Plover on the driest parts of the washes at Purls Bridge all week. A Bar-tailed Godwit passed through on 24th; a Wood Sandpiper turned up on the lagoons on 30th; a Spotted Redshank spent the week with us, and was joined by a second at the end of the month; single Greenshank on 23rd and 27th; Dunlin constant at 40 birds all week. The Ruff are still active, 44 on 25th, the maximum seen on one day, most in good plumage now but still no more than five Reeve.

The lagoons became very dry at the end of the week and Mr Morris, who farms at Purls Bridge, kindly lent us a tractor with irrigation pump over the weekend. Four days of constant pumping on the three lagoons restored the situation, the high tides being too low to do this in the normal way.

The only new summer migrant during the week was a House Martin on 29th. Still Fieldfare with us, though only one seen on 30th. There must be young of Redshank, Lapwing, Snipe and Black-tailed Godwit about by now, but none of us have seen any yet.

MAY

1st week. On the whole it has been a fine week, and dry but for thunderstorms on 6th, clear signs that the washes are drying out. The non-breeding flock of Mute Swans on the lagoons we are maintaining at Purls Bridge has been slowly increasing from 38 to 51 birds throughout the week. Birds are coming in from Earith where it is getting very dry. The non-breeding Coot flock has started to increase from 98 to 124 birds, coming off the drying out washes in the area.

The gull roost, reasonably small by now, is also concentrated in this area and slowly breaking up. Black-headed Gulls dropping from 500 to 300 birds during the week, Herring Gull from six to nil, Common Gull 20 to 12 and Great Black-backed Gull, surprisingly high, from 40 to 19 by the end of the week.

A number of species have now been seen with their first broods. On 6th Shoveler families of ten and nine recently hatched chicks on the lagoons. On the same day we proved Pintail breeding for this year when we noticed a single duck carefully taking two youngsters across the lagoon, keeping close to cover the whole time. We also saw our first Black-tailed Godwits with young on this day, near the Delph river on sand and gravel wash. A family of three minute chicks, very downy, a pale ginger colour with blackish marks splotched on top of the head and in about a dozen places on the body; beak Lapwing-like, the same colour as the enormous legs, which are a slate grey.

The Ruffs have been disappointing, compared with last year, only four left in the Purls Bridge area. All in excellent plumage; two mainly black ones, one with ginger head and black collar; the white one with copper-coloured cap and ginger-mottled mantle. The fourth bird has a sort of parchment-coloured head with purple 'Elizabethan collar'. There are still at least five different Reeves in the area, though by the end of the week we saw little of them.

Some more migrants turned up during the week: a Grasshopper Warbler on Barrier Bank, its song similar to a fishing reel slipping on clutch; a Cuckoo seen flying into willow scrub with its hawk-like flight, and a Swift, all on 1st. The following day a Whinchat on Cradge Bank and a Turtle Dove purring away in the Old House Wood (a call I always associate with hot summer days). And on the day after a Reed Warbler, and on 4th a Lesser Whitethroat. Between 40 and 100 Sand Martins hawking flies on the lagoons during the week with up to 50 Swallows. A lot of Yellow Wagtails about, too, with over 50 feeding round the hooves of the Herefords on Barrier Bank at Welches Dam. A few more cattle on the washes, six between Mepal and Welney now have cattle on them, and one wash with 12 horses and six foals.

Wader passage has been a bit slow, which must be due to the weather because the nice soggy washes are in good condition for waders. Three Green-shank passed through on 1st. Golden Plover numbers dropped from 200 to 60 through the week, a Turnstone spent the night in the washes 2nd/3rd. A late Jack Snipe on 6th and a lone Curlew two days earlier. The Wood Sandpiper which came at the end of last month stayed with us until 5th.

2nd week. The second week in May sees the start of our work to discover how many of each species are breeding, and to collect data on the habitat requirements of the nesting wildfowl and waders. To estimate breeding numbers we plot all sightings of duck and wader pairs on to the maps daily, and with passerine species the singing males. This information slowly builds up to tell us fairly accurately the breeding numbers on the Ouse Washes.

A lot of Shoveler and Mallard broods about. The Mallard have not had a good season, comparatively few broods and these mainly two, three or four. They hatched their broods during the cold weather in April and few survived. On the other hand, Shoveler are doing really well, plenty of broods nine or ten in each; they hatched off over a week later than the Mallard, missing the cold spell, and are now going great guns. Waders too are doing well, large numbers of Lapwing, Snipe, Redshank and Black-tailed Godwits all with good brood sizes.

On 12th we found our first Reeve's nest, with a single egg. The Ruff are much less active now, only displaying in the early morning and late evening, or when one of the Reeves turns up. The poor wader passage this spring could have something to do with it. However all the other breeding waders are having an excellent season. The only passage waders through this week have been Ringed Plover on 9th and 11th, Golden Plover on 9th and 12th and a Curlew on 9th. This week last year we had ten species of wader through, including Little Ringed Plover, Turnstone, Curlew, Spotted Redshank, Green, Common and Wood Sandpipers.

The three lagoons we maintain between Purls Bridge and the railway are proving their worth. At least 30 pairs of Black-tailed Godwit are nesting in the area, coming into the pools in small numbers throughout the day to feed, wash and preen. They really pour in during the late evenings, with much 'wik-a wik-a wik-a' noise, for a communal wash and brush up. The Redshank, Shoveler and Gadwall are also concentrated in these areas.

Good views of an immature female Marsh Harrier working the washes on 12th. Two more migrants in, a Blackcap singing in the willows by the Old House on 8th and a Whitethroat uttering its scratching song from the nettles in the garden on 13th. A very late Fieldfare still with us on 15th, 'chackking' off towards the Old House wood from the washes. The Great Crested Grebe hatched three young on the pool at Purls Bridge on 10th. These were difficult to count because most of the time they were riding tucked away on the back of one of the parent birds.

3rd week. On 15th Gareth, James, Clifford and I did a transect up the centre of the Washes. Clifford and James working from Mepal to Purls Bridge, Gareth and I working Welney to Purls Bridge. We mapped the dominant vegetation of each wash and the state of ditches between each wash, as well as collecting further data on the population sizes of our breeding species. Unfortunately, it was a shade windy so the tendency was to under-record singing passerines.

This walk has given us an excellent idea of our populations. We found 61 pairs of Black-tailed Godwits breeding, and we know two are breeding North of the Welney road; 33 birds were between Purls Bridge and the railway, 10 round Welches Dam, and all but one of the rest between the railway and Welney in two main groups.

For the second year running, Redshank numbers are well up, with 74 pairs mapped; 191 pairs of Snipe were found and 88 pairs of Lapwing. We were only just in time with this transect as many of the young waders were due to fledge soon. We saw no Ruff or Reeve, but did not expect to as the Reeve should all be incubating eggs, and the Ruff leave the Washes to return after the Reeve have finished nesting. I expect that this odd pattern will continue until enough Reeves breed here for there to be always at least one Reeve which has just lost its nest. This would give the Ruff's some inducement to stay.

The wildfowl proved interesting; we found 170 pairs of Mallard, four pairs of Teal and four pairs of Tufted Duck—all down on last year—but Shoveler with 170 pairs and 36 pairs of Gadwall were well up on last year. There was a concentration of duck and waders on the permanent shallow wet areas we have created. All the Pintail were between Purls Bridge and the railway, three ducks and a drake. A little above last year but well down on the year before, a very erratic species this. Garganey showed signs that six pairs might breed and we had two pairs of Shelduck. We saw no sign of Pochard but found four pairs of Wigeon, though we are not yet satisfied that this latter species is going to stay and breed.

All the signs are that Sedge Warbler pairs are 50% up and Reed Warbler 100% up on last year's numbers when Russell mapped the distribution of these species. Our campaign to find the breeding distribution of Yellow Wagtail, Meadow Pipit and Skylark took another stride forward. There was a heavy passage of Swifts moving SW up the Washes all the time we were counting.

On 18th we had two Black Terns and the next day three more. On both days it was in the early morning, and they dropped from high out of the sky, spending a few brief minutes flying over the lagoons we are making to try and get them to nest regularly. They gained height very rapidly on flying off. Conditions on the lagoons are obviously not right yet.

All but one of the Tawny Owls in the Old House Wood had fledged by 19th. They were perched here and there on branches of tall Willows in the vicinity of the nest, and easily found because of much clatter from objecting Song Thrushes and Blackbirds in the area.

Talking of Blackbirds, they have nested in the hide, two nests in fact in the first hide. I think it must be two hens with one male because the nests are no more than seven feet apart. The Wren which nested in the outhouse last year is trying to build under the rear mudguard of my Land Rover again. It scolds angrily each time I get into the vehicle and drive off, and when I return it starts singing loudly.

4th week. The last days of May have been pretty hectic, trying to complete our work on the habitat requirements of the breeding waders and wildfowl with the research department. On 22nd we were all sitting on the Pineapple weed by the gate on Lloyd's wash having completed one bit of work and getting sorted out for the next, when Clifford drew our attention to a big raptor being mobbed by a pair of Black-tailed Godwits and a pair of Lapwing. James sings out 'it's got a forked tail' and sure enough we were looking at a Red Kite. The Godwits looked quite small against it but were pushing home their attacks, diving to within a few inches of the Kite, which seemed not in the least distracted. It flew off towards Ely before our appetites for the spectacle were fully satisfied.

The next day I was doing some typing when Mr Clarke, a farmer from Purls Bridge, came to tell me an owl was caught in his barn. I went back with him and found a Little Owl with one foot that had slipped down a V-shaped gap in the wooden slats forming the side wall of the barn. I lifted it out and found it badly rubbed but otherwise undamaged, cleaned up the wound and released the owl. It was fortunate that the bird had been noticed as it could never have extricated itself. I think it survived, because Clifford and I heard the pair calling a couple of evenings later.

During the last six days of the month we had between 500 and 700 Swifts feeding low over the washes each day. This rather dull cold summer is making the Swifts much more noticeable than usual as they are feeding much lower than in more normal years.

On 25th I had to meet Ted, who shepherds the cattle on our washes, to organise the work of the Hymac mechanical digger that we shall be using to clear out ditches this year. The digger is an important factor in the increase of breeding duck and waders on the Washes. In the old days ditches were cleared by farm labourers, many men working for many hours all over the washes, and giving the birds little opportunity to settle and breed. The digger is much less disruptive, so on this occasion technology is materially assisting in improving our heritage instead of hopelessly polluting it. If we could only organise a reasonable balance between all these things it would be a wonderful world.

The next day Clifford and I walked up Cradge Bank to make our weekly record of the grazing distribution of cattle and other animals on the Washes between Mepal and Welney, at the same time mapping broods of wildfowl and wader chicks seen. We had 1,024 cattle on the washes and 256 on the banks, mainly bullocks, some cows and calves, and a couple of bulls; 30 horses, 13 foals and 500 sheep. During the walk we met Ted again who asked when the Reeve's nest would hatch as he needed to put cattle on that wash as soon as possible. He kindly offered to wait until the nest hatched. I calculated that they should hatch on 2nd June and he agreed to wait until I had seen them hatch out, which was very good of him.

It was a day of squally winds gusting to gale force with some rain showers, causing considerable blowing of the peat soil on those fen fields where the crop was not yet well advanced. At times it was so bad that Manea village was invisible from Purls Bridge, the soil blowing across the Washes like clouds of thick black smoke. The window sills and outside door entrance in the house were deep in black dust on return that evening.

On 30th Gareth, Clifford and I did a centre transect from Berry Fen, Earith to the Mepal road. This is a quiet section of the Washes, but interesting and the narrowest part. It is little grazed and left rough as a result, but there are two fields of wheat with much Amphibious Bistort growing between rows. In some parts we were pushing through Curled Dock and Meadow Sweet five feet tall, but much less Great Water Grass than down our end.

We found 25 pairs of Mallard, only four pairs of Shoveler, two pairs of Mute Swan, five Coot and three Moorhen pairs. Waders were a bit better, but confined to the two ends nearest the road where there are some grazed fields. Redshank six pairs, Lapwing 18 and Snipe 19. Three pairs of Corn Bunting and a pair of Great Spotted Woodpeckers must be the only ones breeding on the Washes. We also had a pair of Grasshopper Warblers, all these near Earith.

In the late evening we were in one of the hides at Purls Bridge when 31 Black-tailed Godwits came in for their ritual evening wash and brush up. At least 18, possibly as many as 22 of these were young of the year. A bit dark to be sure of all of them, but we were pleased to see these because the young are in the habit of leaving the Washes very quickly on fledging; presumably to go to some estuary like the Wash not so very far down the river, which makes it very difficult to assess production.

The month wound up with our first brood of eight Gadwall chicks on one of the lagoons. The chicks were surprisingly large so the ducks must have been very secretive.

JUNE

1st week. We have been carrying on with our transect work down the middle of the Washes. On 1st Gareth and I did the stretch from the Welney road to Welmore Lake. Ivan took us round to Salters Lode and dropped us there, so we could walk to the car Gareth had left back at Welney, thus saving an extra five or six miles of walking.

The washes here are more difficult to work than in the centre stretch, though the general tendency is for a greater rate of grazing, causing the washes to have a shorter sward. The ditches are less convenient, the crossing points being nearer the Delph river, adding an extra half a mile walk for each wash crossed. On the first section of the walk the washes were cultivated, but round the grid pylons they were very undulating, ideal areas for duck production; unfortunately it was all drained and dried up, with comparatively few duck breeding. We then came to the Wildfowl Trust area where much work is being done to construct wet areas. This should eventually provide increasing numbers of breeding duck, providing there are sufficient rough areas to hand, but as yet there were few breeding wildfowl. On the other hand the short sward paid off with waders: considerable numbers of Lapwing, Redshank, and Snipe nesting. Thirty-four pairs of Redshank, 25 with young; 140 pairs of Snipe; 66 pairs of Lapwing, 29 of them with young, and a pair of Oyster-catchers with young. There were no Black-tailed Godwits or Reeve, but we knew that two pairs of Godwits had nested earlier—presumably they had left for the coast soon after fledging.

Ninety-three pairs of Meadow Pipits, 72 of Skylarks and 35 Yellow Wagtail pairs were also mapped. Threequarters of the way through we were caught by rather a severe thunderstorm and had to shelter by a gate for about ten minutes. Luckily we were dressed for the occasion and so avoided a soaking.

The next day we did a second transect from Mepal to Welney, James and Clifford doing Mepal to Purls Bridge, and Gareth and I, Welney to Purls Bridge. So, between us, we completed a centre transect of the whole Ouse Washes in three days. This together with our censusing of the edges, plus a number of complete washes, gives an excellent idea of the breeding numbers of all wildfowl and waders on the whole of the Ouse Washes, with the distribution mapped. The more spectacular birds seen included two Marsh Harriers, a male in the Mepal section and a female in the Welney part; and one of our Reeve nests had recently hatched off, the parent bird circling slowly and silently round us as we passed through the territory.

On 5th Ted phoned in the evening to say he had lost a bullock and would we check the Delph to see if it had fallen in. So Clifford and I took the boat down to look for it. No bullock in the river. It was one of those lovely warm still evenings, so on the way back we popped into the first lagoon hide and had the superb sight of 32 Black-tailed Godwits on the lagoon, 13 juveniles, the rest adults washing, preening and feeding with constant contented 'wika-wika-wika' calls coming from the flock. A lot of Redshank, Snipe and Lapwing too. Quite a lot of duck: 15 Gadwall and a family party with eight half-grown young. Three family parties of Shoveler and 20 or so others loafing about; two drake Teal and a dozen Mallard. A pair of Shelduck at the back of the pool. On our way back up the river it was almost dark; we passed eight groups of Shoveler families with between four and 12 youngsters in each family. A fitting end to a marvellous evening.

2nd week. Spent the bulk of the second week in June trying to work out how many Black-tailed Godwit chicks have got to the flying stage. They are having an excellent year, of that there is no doubt. Most seem to be rearing two chicks and some even have three, large ones, too; whereas usually they only manage to raise one. It could be that the duller weather, causing less evaporation, plus the slight rainfall we have had most days has made the ground wetter and softer, which is more suitable. I don't know.

On 10th the Reeve whose chicks hatched on 2nd stopped escorting them the whole time, spending much of its day feeding on the lagoon—an easy bird to single out, chocolate on its head and a few black splotches on the side of its breast. No wonder Reeves are so hard to follow up; they fly off quietly as one gets near the well hidden nest, and no alarm cry if one gets near the young, which are left to their own devices much of the time after the first few days. None of the usual clues one would expect from a breeding wader.

As 12th last year we had Squacco Heron. James and I were going down the river in the punt in the early morning to do some work, and there it was, this small cream-coloured heron standing at the side of the river in full breeding plumage. It amazed me how small it was, not much bigger than a Lapwing. We were able to drift past in the punt and it stayed at the river's edge. Needless to say, neither of us had our camera handy.

A lot of Lapwing about this week: flocks of 100 or more on the well-grazed washes and small flocks of a score or so birds flying west at intervals throughout each day. They must have done well, elsewhere, as well as on the Ouse Washes. The Black-tailed Godwits are thinning out now as more and more chicks reach the flying stage and drift off, presumably to the nearest estuaries. The drakes are beginning to go into eclipse and becoming more secretive and less active, which is understandable since they are flightless during the moult of main flight feathers. This week has seen a big drop in Coot numbers. The last of the non-breeding Swans left us on 10th. I wonder where they go to moult? Whilst recording cattle, Clifford had a family of Long-tailed Tits near Mepal, including four newly-fledged young, so at least one pair has bred successfully on the Washes this year.

3rd week. The third week in June sees our breeding waders and wildfowl well advanced; just the re-nest waders still busy with young, the rest finished and drifting away. This is a good time for seeing the birds on the lagoons as they have time to spare for loafing about. Most of the duck have finished nesting, but there are a few late broods of Mallard and Shoveler, and Gadwall broods are well advanced. Tufted Duck and Garganey are still sitting on eggs or have recently hatched chicks. Those drakes still on the washes are fast going into eclipse plumage.

At the beginning of the week Gareth collected samples of the life in the lagoons we maintain—areas where the ditches are dammed to allow water to flood over onto the washes, causing conditions as near to summer flood as we can manage. The main problem with these lagoons is the taller vegetation, mainly Great Water Grass, growing up and smothering the water area, so we have to remove this to leave large open water areas. Gareth found vast numbers of midge larvae, newts, sticklebacks, etc., in these lagoons; food for the waders and wildfowl.

On 18th, while collecting records of cattle grazing weights on the Washes, we saw a stoat carrying off a rat. We are lucky on this reserve, there is an excellent balance of mammals because the winter floods result in a very low spring population of mammals per acre. We do not control Stoat or Weasel because they contain that rat populations better than we could, but as there are comparatively few stoat or weasel they do no significant damage to our breeding birds, since none of the important species are vulnerable ground colonial nesters.

On the walk we found a pair of Black-tailed Godwits with two chicks on

Cradge Bank and herded them back into the nearest wash. The adult birds stooped to within inches of us and gave distraction displays whilst we carefully persuaded the chicks into the wash. They are much safer there than on Cradge Bank where there is much coming and going of farmers and shepherds with their dogs and vehicles.

4th week. There is still a flock of 30 Black-tailed Godwits, which have finished nesting, feeding on a well-grazed wash near the lagoons. Frequently coming and going between the wash and lagoons to bathe, drink and take a change in diet. Groups of juvenile Godwits are coming into the lagoons in the evenings before leaving the Washes. Reeve activity is very slight, an occasional bird to be found on the lagoons most days. Our first juvenile Ruff came to the pools on 25th. I think only five Reeve have nested, but at least these should be as successful as the other waders.

We spent the week completing the maps of breeding passerines and collecting more data on duck and wader brood sizes. We had a spell when Sedge Warbler had newly hatched young and stopped singing during the day, making censusing difficult, but they have started again now. Some of our visitors express difficulty identifying Sedge Warbler song from Reed Warbler. The best way I know of distinguishing them is to notice how irritated a Sedge Warbler sounds with life as it 'cherrs' away, whereas a Reed Warbler has a most contented type of babble, reminiscent of quiet summer days.

The Great Crested Grebes chick on Purls Bridge pool is quite big now, but I doubt if it can fly yet. We have not seen the Shelduck since 23rd—they have probably taken their chicks into the tidal New Bedford and got them off to the coast without our seeing them. An immature Marsh Harrier was quartering the washes on 29th; we have seen less of them so far this year than last. Lapwing continue to fly west in small groups, masses must have passed over by now.

Two juvenile Tawny Owls still frequent the Old House wood, and I found a Barn Owl's nest near the house on the last evening of the month. While phoning Sandy in the evening I saw a Barn Owl take three small mammals past the window in as many minutes, so on completing the call I went out to look for it. There it was right under my nose, two eggs in the top of a pollarded Willow. No wonder our searches of all the local barns only produced bags upon bags of owl pellets.

JULY

1st week. The month started well with a Bittern flying into the second hide lagoon, the first Bittern we have had in summer since I came here. A Water Rail came out and showed itself on the same day.

The weather has been close and unsettled all week. Ted became ill near the end of the week and phoned through to ask us to check on a portion of the cattle for him until he recovered. Counting the cattle added considerably to our work each day, but it was an interesting experience and we found two broods of Teal in the ditches we would never, otherwise, have known about. Even so, each day brought new problems: a very misty morning on 6th added to our difficulties, and the next day a herd broke through one of the stowaway fences to join the cattle in the next wash. After we had sorted out the new count of cattle in each wash we had to collect posts and barbed wire to renew the fencing, leaving less time than ever for our own work.

The number of Mallard and Gadwall have dropped steadily throughout the week. Very few ducks of any species to be found that are not well into eclipse now. Two juvenile Pintail came in on 4th; a lot of Shoveler on the lagoons, mostly juveniles. Shoveler are by far the commonest duck at present.

An immature male Marsh Harrier with very tatty primaries, being well into moult, arrived on 4th. We had quite good views of it as it searched up and down the ditches with lazy flight. The lagoons are persuading some of the newly fledged Black-tailed Godwits to stay. In all previous years they have left within a couple of days of fledging, but now we are coming to expect at least 10 birds on the lagoons. Two Reeves and a Ruff came in on 2nd but we have not seen them since. The Grasshopper Warbler by the first hide has started singing again. I wonder if it bred near there?

2nd week. I suppose this is our quietest time of the year for birds. Almost all the activity that can be seen is on the lagoons. Most of what happens on the Washes goes unnoticed; the grass so long at this time of the year that it grows more quickly than the cattle can eat it. A drake Wigeon dropped into the lagoons on 14th; two Pintail, a drake and a juvenile, spent 13th and 14th with us. Teal numbers have built up to about a dozen birds; only 40 Mallard and half a dozen Gadwall, all in eclipse, but still good numbers of Shoveler. A Tufted Duck has a brood of four chicks in the Delph between the lagoons.

There are indications that wader passage this autumn may be better than last year. A Greenshank and a Dunlin on 8th, a Curlew on 11th, and a Golden Plover on 14th, all spent part of the day on the lagoons, the only wet areas left on the marshes. Still 20 or so Redshank, about 50 Lapwing, a dozen Black-tailed Godwits, and at least 100 Snipe, these last are very hard to see in the wet areas amongst the Amphibious Bistort at the edge of the lagoons. Ruff and Reeve to be seen every day this week, though only in ones and twos.

Some signs that autumn is on the way; a juvenile Great-spotted Woodpecker spent the week in the tall Willows between the hides; up to a dozen juvenile Sand Martins were flying over the lagoons at the end of the week; a Kingfisher flew up the Delph on 8th.

Counting cattle on 9th took a long time as a heifer had fallen into a ditch and it took us two hours to ease it out with the Land Rover, extra care being needed as there were a lot of old posts in the area, and we had to be sure not to pull it over these. However we found a Barn Owl's nest in a pollarded Willow on the Washes near Welney on our way back at the end of the day, so the delay was worth it.

3rd week. I was very glad for Ted that he had recovered from his illness by 15th; good for us too as we need the time for preparing the lagoons for next year. The Great Water Grass has to be cleared otherwise they soon become overgrown and useless for waders or wildfowl.

We found two pairs of Whinchats, complete with their newly fledged broods, on the washes near Cradge Bank opposite Oxlode on 15th, and had excellent views of three Marsh Harriers, two males and a female, which remained with us for the rest of the week. The first Kestrel of the autumn came in on the same day.

Sand Martin numbers have been building up all week, until by the end of the week we had a good 200 with us. Going down the river on the Friday, we found a Sand Martin swinging from a Willow overhanging the river, its beak caught in a fish hook which had tangled in the tree. Fortunately, we were able to stop the boat under the branch and release the bird, which flew off apparently none the worst for the experience.

The Great Crested Grebe chick on Purls Bridge pool finally flew off on 20th. We also had two Common Terns flying down the Delph on that day, as well as a Short-eared Owl which also stayed the next day. After a long absence three Garganey returned to the lagoon on 17th; they are easily confused with Teal when in eclipse as some have only a faint buffish-white eye stripe to identify them by. A further difficulty is the fact that some Teal may show a similar stripe, though always much smaller. In flight, identification is much easier, and the white underwing with black leading edge is a give-away even if the speculum has moulted. At the end of the week a Pochard with four well-grown chicks swam cagily across the lagoon. Our first proof of breeding on the Washes this year, and quite unsuspected.

4th week. This last week of July has been quite good for waders on the lagoons. There must have been at least 200 Snipe; ten juvenile Black-tailed Godwits, and an adult which had all the characteristics of the Icelandic race spent 26th and 27th with us. Redshank steadily dwindled until by 30th none were to be found, but up to five Ruff and Reeve were present all week. Single, Common, Green and Wood Sandpipers passed through the pools during the last five days of the month, mostly Wood Sandpipers; a Greenshank on 30th and a Dunlin on the last three days of the month. Still plenty of Lapwing about.

We spotted two Little Terns feeding in the Delph river on 22nd and spent minutes watching their dainty flight as they fished up and down the Delph, always working into the wind. Eventually they flew off, eastwards.

We have amused ourselves by watching the juvenile Herons which spend hours catching dead Great Water Grass rhizomes on the lagoons. They must confuse them with eels. Once one did catch an eel, and took it on to the soft mud to eat but made the mistake of putting it down to get a better hold of it, and spent the next 20 minutes trying to figure out where the eel had gone. A hard life, learning to find your own food so young.

Only two Marsh Harriers seen this week, both males, but kestrels increased from three on 29th, to six by 31st.

AUGUST

1st week. The Washes look very ragged at this time of the year. Seen from Purls Bridge the Barrier Bank is an untidy mass of seeding Thistles; there are Teasel, too, their pale lilac-pink flowers encircling the flowerheads like a monks hair-do. Large charms of Goldfinches fly from the banks to drink at the edge of the river and back to feed again. The Osiers are in full leaf along the river, a screen of green choked at ground level with *Convolvulus*, Dock, Purple Loosestrife and Comfrey.

Across the river each wash looks different. The nearest, well-grazed near the river, is stifled with Thistles further back, and a herd of 40 mixed cattle graze quietly in the middle stretches. The adjoining wash is ungrazed at the moment, rough with Great Water Grass and overgrown Tufted Hair Grass. A battered old gate rests against crazily leaning gateposts, and a scruffy Heron, heavily in moult, stands humpbacked and rock still on the right hand post. The third wash is a neat narrow belt of smooth green, well grazed by a herd of 40 bullocks, followed by another of rusty, ripening Curled Dock, lightly grazed by 20 heifers. The next a yellow mass of flowering Marsh Ragwort, and away into the hazy distance regular bands of colour and texture show the effect of different grazing pressures on successive washes.

There are hundreds of Sand Martins, now, flying over the washes and concentrating close to the cattle. An occasional Yellow Wagtail, Reed Bunting and Meadow Pipit flies to and fro across the river, Moorhens skulk at the rivers edge and a family of Mute Swans with five young goes swimming proudly down the river. Away in the distance a Marsh Harrier can be seen quartering the fields; flocks of a score of Mallard, Teal or Shoveler rise in panic if it goes too near. Panning the sky, eight Kestrels can be counted all hovering in the still air.

On 4th, down at the lagoons, there are still four juvenile Black-tailed God-wits, all feeding in deep water. Redshank have dwindled to three birds; only one Lapwing, now, the rest in the grazed washes, but still 50 Snipe to be found. There are four Ruff round the edges of the lagoons and, lurking in the Amphibious Bistort, a Wood Sandpiper with beyond it Green Sandpiper feeding on the mud. Their calls are quite distinctive, the Wood Sandpiper's like a penny whistle; the Green Sandpiper's clearer, like a clarinet. Over the river, Broadbodied Libellula dragonflies with powder blue abdomens flit among Shoveler duck, and a Grasshopper Warbler reels away on the opposite bank. This is a bird which probably bred there, but we don't know for sure.

2nd week. A dry warm week, the water on the lagoons is at a very satisfactory depth for waders. Much to our surprise the Black-tailed Godwits increased from six to nine birds during the week. Gentle wader passage starting with a Dunlin on 9th; two Greenshank coming in next day, that gorgeous clear 'too-too-too-too' note ringing out across the Washes. Two days later quite a rush with two Curlew, a Green Sandpiper, eight Wood Sandpiper and a Common Sandpiper, all on the same day. Two Wheatear also arrived on the banks so there must have been a general movement on this day. A dozen Ruff now and at least 200 Snipe but no Redshank.

Swift numbers dropped from 60 to 10 during the week but Sand Martin have increased, there must be a good 4,000 now. Warblers are on the move, we still have plenty of Reed and Sedge Warblers in the Osiers, but some have obviously left. Willow Warblers are passing through in ones and twos each day. Their 'seep' note giving them away as they flit about in the Osiers. The ducks being all in eclipse are now very quiet, only two score Mallard, a dozen Teal, eight Garganey and 30 Shoveler on the lagoons these days.

3rd week. Duck numbers have been dropping all week, Mallard down to 20, although Teal have remained about the same; but only three Garganey around and two Gadwall; even Shoveler are reduced to a score of birds.

The waders are poor, too. The lagoons are very low, but nothing can be done about it until the full-moon tides next week. This state of affairs is probably the reason for a mere two Black-tailed Godwits, and Ruff down to six birds. Snipe is the only bird to hold its numbers. Wood Sandpiper lasted to 18th and a Dunlin stayed with us until 19th; three Golden Plover flew through on 17th, a Curlew on 19th, and two Greenshank on 21st, but none dropped in.

Sand Martin numbers have increased to 10,000 but they are not roosting on the Washes—not as in '68, the vintage autumn, when we had a roost of between two and five million between Purls Bridge and Welches Dam, a three-quarter mile long Osier bed crammed full of Sand Martins. On windless fine evenings the sky darkened as they flew in four huge smoke-like cones above the roost. Countless thousands in each cone, the bottom of the cones sweeping backwards and forwards over the Osiers with hundreds of Sand Martins dropping out into the Willows from the bottom as the cone passed over the Osiers. The noise was outstanding, the chatter like a vast waterfall; an unforgettable experience I may never have the chance to hear again. During the day the Washes were covered with Sand Martin. Every Dock on every wash had birds perched on it and huge numbers flying about. The roads were covered in a carpet of Sand Martins sunning themselves and, it was impossible for vehicles to pass without killing a few. I shall never forget it. Good migration conditions to the North, bad to the South, and the flooded washes, were the probable explanation.

4th week. At last some ducks have started to come in: 300 Mallard and 50 Teal arriving on 29th to boost our dwindled stock, and a further 200 Teal on 31st. Garganey have dropped from five on 24th to only one by 29th, and Shoveler from eight on 25th to none at the end of the month. Three Wigeon came in on 30th, all immaculate drakes, followed by 50 more next day.

The frustration of the week was a probable Temminck's Stint which we saw feeding on a small patch of soft mud at the back of the lagoon on 30th. We could only view it at long range and were unable to see all of its plumage, nor did we see it fly or hear it call. So brackets will remain round the record. We were awkwardly placed but to have walked over to flush it would have been an unwarranted disturbance to other birds.

Black-tailed Godwits lasted on the lagoons until 27th. Ruff dropped from eight birds on 22nd to a single bird at the month end. Greenshank, Common Sandpiper and Curlew were passing through in small numbers throughout the period. The lagoons were back to a respectable depth of water by 27th, and there were three Swift still with us on 31st.

We found a Poplar Hawk Moth caterpillar in the Osiers by the second hide, munching through a whole willow leaf every ten minutes, eating up one side, then the other, and finally down the stalk. The caterpillar was at least two inches long and quarter-inch in diameter, looking just like another willow leaf and very hard to see even when one knew where to look. The best way of finding them is to look for a row of missing leaves.

SEPTEMBER

1st week. The 1st September brings in the wildfowling season. This year, with not many duck about, its a comparatively quiet first day. There were 88 wildfowlers out between Mepal and Welmore, and 134 ducks were shot. Some areas had reasonable sport, others very poor.

As in all things there is a variety of people out shooting, from the 'respectable' wildfowler who identifies his quarry, has some idea of range, respects his boundaries and other people, and realises his privilege. These people are most helpful to us on the Washes, cooperating to the full with our research work and, in the main, supporting what we are trying to do. At the other extreme we have the 'marsh cowboy' to contend with, our enemy and the wildfowler's, too. No respecter of boundaries, unable to identify quarry (uncaring), little idea of range, and no knowledge of behaviour. I had to deal with a pair shooting on the bank where no shooting is permitted. The ducks being few, they had nothing better to do than shoot their old cartridge boxes. Thank goodness such types are a minority.

On the whole a quiet week with few waders. A Greenshank passed through on 3rd, another with a Ringed Plover the next day and a Green Sandpiper on 6th. A couple of Swift about until 7th, since then we have seen none. Cuckoo and Turtle Dove are still comparatively abundant in the Willow holts by the river. Swallows have increased to about 1,000 birds, and the House Martins nesting on the Ship Inn at Purls Bridge still have young in the nest. Plenty of Sedge, Reed and Willow Warblers, a Chiffchaff on 7th, and Spotted Flycatcher on the railway embankment. A good 200 Goldfinches on the Thistles on the Barrier Bank with a score of Linnets, and three Redpolls resident in the garden through the week.

2nd week. The weather has become more unsettled after the fine spell of the last four weeks—the new moon I expect. Duck numbers are still building up very slowly indeed, a reflection of the poor Mallard breeding season? I doubt if there are 250 Wigeon on the whole of the Washes, most of these are on the Wildfowl Trust refuge. There are still three Marsh Harriers with us, two males and a female, and it is possible to find up to 12 Kestrels hovering over the area of the RSPB refuge. Very few waders present except Snipe and Lapwing; two Curlew on 8th, a Common Sandpiper and a Dunlin came on the lagoons on 10th but both left on 14th. Eight Golden Plover dropped in on 11th.

We have not seen any Swift this week. The Turtle Dove are still here, but no Cuckoo since 11th and Sand Martin are down in the hundreds now. Plenty of Warblers including a Lesser Whitethroat and Chiffchaff; also a Goldcrest, on 8th and 12th, which is not a regular bird here. Yellow Wagtail are still about, but no winter passerine migrants have appeared as yet.

3rd week. Our first full count of the winter on 15th; James, Gareth, Ivan and I counting. Not a lot about yet: 69 Mute Swans, only 3,800 duck altogether, 2,900 of them Mallard, 500 Teal and 350 Wigeon, only five Gadwall and 20 Shoveler. Wader numbers, like those for duck, are low: 21 Golden Plover, 470 Lapwing, a single Green Sandpiper, Common Sandpiper, Curlew and Jack Snipe, plus 31 Ruff and 140 Snipe. There are still three Marsh Harriers; fortunately, we are careful about flying birds because two of the Harriers flew over different sections during counting and we might otherwise have recorded too many. Eighteen Kestrels counted which is an encouraging score. Still 50 Yellow Wagtails, 2 Whinchats, a Spotted Flycatcher, 16 Reed Warblers and five Sedge Warblers with us. Also single Lesser Whitethroat, Whitethroat, Blackcap and Willow Warbler. We recorded 970 Goldfinch, so at least the thistles have some use.

The next day was a soaker, almost two inches of rain out of the rain gauge, so we were lucky with the count. A flood started on 18th with the water, at 100 cm (3 ft 3 in), beginning to seep on to the washes, and by the end of the week it was a good halfway across at 162 cm (5 ft 4 in), fortunately, it then stopped raining, just in time before panic stations and 'all cattle off'. There were some signs at the end of the week that Duck numbers are on the increase, though it may be that duck are moving up and down the Washes, looking at newly flooded areas.

A Spotted Redshank came in on 20th, and next day we had a Redstart, two Twite and a Willow Tit. The latter fortunately giving its nasel 'cher-cher' call to confirm our field identification, because the Willow is difficult to tell from Marsh Tit by plumage. Both have very similar markings, but the Willow Tit is not quite so grey and has slight pale edgings to the secondaries (which can be abraded away at times); but the light can be so confusing, causing the dull black cap on the Willow to look glossy like a Marsh, or vice versa.

4th week. The flood hovered round the 162 cm mark until 29th when it started to go down, and by the end of the month it was down to 148 cm (4 ft 10 in). The flood has pulled in some wildfowl and waders, but nothing on the scale of the '68 flood. Wigeon numbers have increased to 1,200 and already Mallard are over the 3,000 mark. Surprisingly there are only 500 Teal, whereas in the '68 flood we had over 4,500; however it was much deeper then and lasted much longer. Shoveler have not reacted very much, increasing to only about 100 birds so far. In '68 we had 1,500, but that might have been a bit earlier.

The passage waders are small in numbers, but they had increased noticeably before the flood. Two Dunlin at Purls Bridge, 30 Ruff commuting up and down the Washes, and Spotted Redshank daily with three on 27th, their 'tu-whit' whistle advertising their presence, and two more on 29th. Greenshank present on most days with two on the last day of the month. Surprisingly there were no Wood or Green Sandpipers this week—the missing ingredient must have been the lack of east winds. In '68 on one memorable day we counted 77 Wood Sandpipers, 26 Green Sandpipers and two dozen Spotted Redshanks. However, we have had nothing on that scale since.

Summer migrants are becoming scarcer; we have not seen a Sedge Warbler this week though Reed Warbler are still here. A Whinchat on 24th may be the last of the year, and no Turtle Dove have been seen since two on that day. No Cuckoo, Spotted Flycatcher or Yellow Wagtail, though we had a Grey Wagtail on 30th, pulled in by the flood no doubt. We still have a lot of Swallows, as well as a few House Martin and Sand Martin. The Willow Tit stayed with us until 25th, and we also had a Coal Tit on 2nd and 23rd.

WETLAND   5   One of the flooded washes in autumn, the Barrier Bank in the background. *Photo : Peter Merrin, RSPB.*

6   Cattle grazing the washes in summer, at Purls Bridge,

WETLAND

7   One of the drainage ditches
—the washes in spring. *Photo :
Ackroyd Photography, RSPB.*

8   Winter on the Ouse Washes.
*Photo: Ackroyd Photography,
RSPB.*

OCTOBER

1st week. The water from the sudden September flood is going down quickly from 140 cm (4 ft 7 in) at the beginning of the week to 97 cm (3 ft 2 in) between the Delph banks by the end of the week. Mallard numbers are fairly normal, about 500 on the RSPB refuge, flighting in and out between the stubbles and the Washes. Teal have seemed to favour the washes to the north of the Old House; a flock of 200 were among the tussocks of Tufted Hair Grass. Between 10 and 29 Gadwall in the area on the Lagoons; the third ditch up from the railway seems to be a favourite spot of theirs. Wigeon have been building up throughout the week from 760 to 1,200, in two main groups in well grazed areas on the refuges. Sixty-one Pintail on 2nd, and 21 on 6th, were the only days we saw any, which is as erratic as we have come to expect. There are about 150 Shoveler around. A Garganey on 2nd was a surprise, a drake coming out of eclipse; 47 Tufted Duck and 83 Pochard at the beginning of the week, were down to 24 and 30 respectively—the water, dropping, is doubtless the explanation.

An excellent week for Rails on the lagoons. We have seen between two and five Water Rails daily, and saw a Spotted Crake from 1st to 3rd—close views, but it never called. Quite unusual to see but not to hear this species, as it is usually the other way round. Light but steady wader passage throughout the week: single Spotted Redshank, Redshank and Oystercatcher, four Ruff and 100 Lapwing on 1st; on 3rd two Spotted Redshank; six Ruff and 300 Lapwing next day. By 5th single Dunlin, Redshank and Wood Sandpiper, together with three Ruff, five Spotted Redshank, two Oystercatcher and 400 Lapwing. The numbers increased to 10 Ruff and six Spotted Redshank the following day with single Redshank, Greenshank and Ringed Plover; also a heavy influx of over 1,300 Lapwing counted. A single Redshank, a Ringed Plover, two Oystercatcher and still over 1,000 Lapwing with us at the end of the week.

Two Short-eared Owls on the banks, still plenty of Swallows, and six House Martins dropped in over Purls Bridge pool for a spell on 2nd; a Sand Martin on 4th too. We still had two Reed Warblers on 7th, a Willow Warbler on 4th and a Chiffchaff on 6th. Three Fieldfares came in on 2nd and a heavy thrush passage on 4th, mainly Blackbirds, over 100 counted, but with them 20 Redwings and six Ring Ouzels, the latter not at all common here. Coal Tit and Siskin on 7th; Willow Tit on 2nd and 3rd; all in the Osier beds, and all birds we don't often see.

2nd week. The water level dropped from 94 cm (3 ft 1 in) to 62 cm (2 ft) during the week. The wet areas on the washes had largely dried up by the end of the week but, on the whole, wildfowl numbers stayed constant except Wigeon which dropped to 800, and Shoveler from 350 birds at the beginning of the week to only 70 at the end.

There were considerable numbers of Lapwing going west, plus over 1,000 birds on the Washes, during the first days of the week, dropping to 300 by the end of the week. Wader passage in general has been thinning out with the drying up of the washes. A Curlew Sandpiper on 9th was interesting, two Oystercatchers on 11th and a Greenshank on 14th. Ruff numbers fell from 18 to eight during the week. A Jack Snipe spent the week feeding in full view most of the time on a muddy area on one of the lagoons.

On Tuesday I was watching a Little Grebe in the Old Bedford River north of the railway, when it suddenly flew downstream. A split second later a large Pike rose out of the water just where it had been: a narrow escape.

There is still no shortage of Swallows, and we had 12 House Martins on 9th, two on 11th and one on 13th. A Yellow Wagtail was seen on 11th and a Reed Warbler in the Purls Bridge Osiers on 13th. Collecting the last sightings of the season for migrants is much more difficult than first sightings. Two Bramblings on 13th passed over uttering their rather harsh flight call. Redwing had built up to over 100 by the end of the week, but Fieldfare have been very few. They are late this year.

3rd week. On 15th we did another full count of the Washes. James, Brian, Gareth and I counting, each taking our usual piece. The Delph was at normal level, 65 cm (2 ft 2 in), with little water on the washes other than permanent pools and the Wildfowl Trust's flooded areas. A useful count but not spectacular. Most of the wildfowl on the Trust's refuge: 74 Mute Swans, 34 of them immature, 450 Teal, 4,200 Mallard, four Pintail, 63 Shoveler, two Gadwall, 2,000 Wigeon, 10 Tufted Duck, 25 Pochard and only 38 Coot.

I saw an Arctic Skua being mobbed by gulls along the Delph river and a Wheatear on Cradge Bank; James had a late Turtle Dove near Mepal. We recorded 41 Golden Plover, most up at Earith, but only 231 Lapwing, the large passage of the last two weeks seems spent now. Four Ruff on the Trust's flooded area, over 150 Snipe, and a Jack Snipe near Earith; 417 Skylarks, 280 Meadow Pipits, 133 Reed Buntings, 281 Goldfinches and 96 Linnets, mostly on Barrier Bank between Welches Dam and Purls Bridge where the Thistles have not been sprayed. We recorded 107 Tree Sparrows and 7,300 Starlings.

Among winter migrants seen were a Stonechat near Purls Bridge, seven Bramblings and 158 Redwings, but no Fieldfares. We counted 55 Swallows and eight House Martins, and had 33 Kestrels between us, all but one of the last hovering over the high banks between Mepal and Welmore Lake, no doubt after some of the very high small mammal populations present at this time of the year. Sixty-eight species recorded during the day.

There are still 1,500 head of cattle on the washes plus 230 on the banks; also 400 sheep, 170 horses, 20 goats and 3 donkeys. Most of these will be removed by the end of the month.

4th week. This last week of October brought real signs of the winter species coming in. On 23rd a family party of Whooper Swans came through, four adults and three youngsters. They landed at Purls Bridge in the early morning, with much trumpeting, and spent the rest of the day there, but were gone next day. On 25th two adult Whooper Swans arrived and were still with us at month end. The first Bewick's Swans came on 30th, three adults and thre youngsters; they too dropped in at Purls Bridge and have stayed with us.

On the last day of the month a male Hen Harrier was quartering the refuge —superb plumage grey, black and white—and really stirring up the duck. We still have good numbers of Kestrel on the banks, and the Short-eared Owls are beginning to increase with six between Purls Bridge and the railway; also a Barn Owl and two Little Owls working this section of the banks.

Lapwing numbers continue to decrease from 200 to only 40 by the end of the month, but plenty of Snipe, about 300 on the RSPB refuge, and two Jack Snipe showed themselves on 23rd. Swallows are still with us but only two seen on 31st; the two Reed Warblers seen on 23rd must be the last. Some of the birds passing through must be on passage: a Treecreeper on 25th and two Goldcrests on 23rd. Fieldfare are at last coming in with five on 24th and 20 on 30th; really late this winter.

Four Water Rails on the lagoons, a Great Crested Grebe and a Little Grebe on the Delph, plus a Kingfisher to be seen buzzing up and down, mostly by the Osiers that overhang the river. We estimate that 7,000 Starlings are roosting in the Osiers opposite the Old House.

Plenty of Herons about, a good score or so to be found on the Purls Bridge refuge most days. Duck are building up slowly, a few more Pintail, 50 on the RSPB refuge on 29th. The refuge is disturbed weekends by wildfowlers, who own washes near the middle of it, and shooting sadly splits up swan family parties; fortunately the Wildfowl Trust refuge is more secure. Most wildfowlers are keen on refuge areas and can see the need for them. Some have been very cooperative and with conservation in mind rather than their own convenience have helpfully moved away from the refuge area.

NOVEMBER

1st week. Mild dull conditions most of the week except 6th when we had heavy rain, 10 mm (2/5th in) in the rain gauge. The Delph river has been between 60 and 70 cm (2 ft to 2 ft 4 in) a highish normal level.

At the end of the week we started to put the foundations down for two bird hides on Barrier Bank at Purls Bridge, a slow job at this stage. The foundations have to be level, so each hole for the legs has to be dug accurately in relation to the others, both in position and depth.

The wildfowl numbers remained similar to last week: Bewick's Swans at Purls Bridge still six birds, but the Whooper Swans had increased from three adults to six adults and a youngster by the end of the week. The Hen Harrier stayed with us until 2nd, and we could easily count up to 7 Kestrels hovering over the Barrier and Cradge Banks. Only one Short-eared Owl, though we always seem to notice less near the full moon; I suppose its easier for them to hunt at night during a full moon. Plenty of Little Owls at Welches Dam, Purls Bridge and the tall trees near the railway.

Some movement of birds passing through, specially on 6th when we had a Greenshank, the loud clear 'tu-tu-tu-tu' call ringing across the Washes as it flew down towards Denver. A Bearded Tit in the Osiers at Purls Bridge which, as so often with this species, gave itself away by its loud 'pinging' call as it climbed to the top of an Osier. Finally a Twite flying across the Washes at Welches Dam, identified by its flight call which is similar to the Linnet's until it utters the phrase that sounds like 'twite'. In the winter months, Twite can be found in small flocks on the farmland surrounding the Washes without difficulty if one knows the bird, but if not it is easily confused with the even more numerous Linnet. It is also a tide-line feeder on the flooded washes during most winters.

181

A noticeable amount of movement from Long-tailed Tits throughout the week. Flocks of 11 on 2nd and 15 next day, working through the Osiers between Welches Dam and Purls Bridge. Five Stonechats are now about between Purls Bridge and the Railway on Barrier Bank; I like to have this species around as it adds that extra something to the place. Probably the main interest of the week has been the Great Grey Shrike which turned up on 3rd and, very obligingly, has been present near Welches Dam ever since.

The Starling roost in the Osier bed opposite Colony is having a rough time because workmen from the Great Ouse River Authority are cutting the Osiers. They weave them into mats to fix to the river banks in order to consolidate them, and the roost habitat is shrinking as a result. The cutting must be messy with the smell and deposits left by the roosting birds.

2nd week. Still no flood, despite the rain last week. We put one of the hides up on 9th, Ivan, Brian and I. It was windy when we started and blowing a gale before we finished. At one stage I had to hold the roof down, hanging onto a clamp fixed to one of the main roof beams, and was blown off my feet on more than one occasion. We managed to get the building up in the end and by the time we had finished we had proof it would stand up to any normal gale. Ivan and I put up the second hide on our own; there was no wind and it went like a charm, no trouble or effort compared to the other.

We saw the last of the Great Grey Shrike on 8th. Our sightings of Short-eared Owls increased as the moon waned from full. None on 10th, one next day, two on 12th and 13th, and eight by 14th. Still plenty of Kestrels, and a female Hen Harrier spent much of 12th flying low up and down the washes opposite Purls Bridge, stirring up the wildfowl.

No dramatic build-up in the wildfowl numbers yet but Wigeon are starting to increase significantly now. About 1,500 birds on the RSPB refuge. We have not put any water down yet, and until we do will not pull many in. The swan population remains more or less static, but six more Bewick's Swans came in at the end of the week, and seven Greylag Geese on the same day. These Greylag are certainly feral birds.

Not many waders: about 50 Lapwing around and plenty of Snipe: one can find a hundred of these without much difficulty. Four Golden Plover and a Jack Snipe on 10th. We underrecord Jack Snipe, unfortunately, but I see no way round this, since an accurate assessment would cause too much disturbance to the wildfowl. Three Ruff by the lagoons next day and two Water Rails.

The gull roost is still low, only 100 Black-headed Gulls, six Common Gulls, 30 or so Herring Gulls and six Great Black-backed Gulls, but I expect the Wildfowl Trust have plenty as they have shallow-flooded their washes already. The Starlings in the Osiers at Colony were finally driven out during the week when the Osier bed became too small for them. I do not know where the new roost is, somewhere out towards Littleport I should think.

3rd week. One way and another this has been quite a hectic week. James, Gareth, Brian and I did a full count of the Washes on 19th; water level only 66 cm (2 ft 6 in), the weather cold with a north-west wind blowing, and snow falling in the morning after a frosty night. The snow became so bad that we had to stop counting until it finished at about half-past eleven. The wildfowl are now building up well: 95 Mute Swans, 25 of them immature, nine Whooper Swans, all adults, and 79 Bewick's Swans, 12 of them immature. Most of the Swans are on the Wildfowl Trust washes. The only geese are the seven Greylag still up at Purls Bridge. We counted 623 Mallard on the Washes, but a lot are in the stubble and old potato fields on the Fens; 358 Teal, two Pintail and 59 Shoveler; Wigeon were up to 7,000. The only diving duck present were 21 Tufted Duck and three Pochard. Coot were low at 85 but Moorhen at 300 were about average.

Wader numbers were interesting, with 203 Golden Plover, 2,500 Lapwing, only one Ruff, a Jack Snipe and 250 Snipe. Other birds included a Great Crested Grebe at Mepal, a Little Grebe near Welney, a Black Swan in the Delph near Welmore Lake, 31 Herons, 18 Kestrels, and 18 Short-eared Owls. James had six Kittiwakes flying up the New Bedford River. In all, 62 species recorded during the day. There were still 740 cattle, 170 horses, 190 sheep, 30 goats and two donkeys on the Washes.

The next day we had more snow and spent most of the time damming the ditches and centre wash grips at the Delph end of the washes, at Purls Bridge, ready to keep shallow water for wildfowl on the refuge. The snow reminded us of the previous year when we had a huge passage of Skylark flying west on a very broad front after a blizzard at this time of the year. We estimated that 14,000 went over, so goodness knows how many birds were involved in the movement.

On 21st we noticed the river Delph changing colour and starting to flow, and by evening the level was up to 90 cm (2 ft 11 in) and beginning to spill over the banks onto the washes. The dams had been in the nick of time. There must have been heavy rain in the Midlands at the beginning of the week because there has been insufficient here to cause the flood. A female Hen Harrier was quartering the washes as we watched the river rising.

4th week. At last the proper winter floods seem to have started, with the water at 114 cm (3 ft 9 in) spilling onto the washes on the morning of 22nd, and rising to 149 cm (4 ft 10 in) halfway across the washes by 26th. The level had fallen to 140 cm (4 ft 7 in) by 28th, rising again to 156 cm (5 ft 1 in) on 30th after more heavy rain on 28th.

This flood really has started things moving: Mallard up to 500 on the RSPB refuge, Teal up from 100 to 200, Wigeon doubled, and 60 Pintail coming in on 27th. Tufted Duck went up from seven to 20, and Pochard showed interest on 25th when 30 birds arrived. The seven Greylag have remained with us—they must be feral birds because they ignore the activity of nearby wildflowers who can only look at them in frustration.

Bewick's Swans have been coming in every day at Purls Bridge with 48 on 22nd, up to 163 by 26th but only eight juveniles, and 252 on 29th with 20 young. The Whooper Swans have remained at nine birds, five of them immature; the Mute Swans have increased from 12 to 20 birds.

The flood waters really bring the Washes alive, even Coot are noticeable now, over 200 at Purls Bridge and the gulls, too, have piled in since the flood. Black-headed Gulls have risen from 400 on 22nd to 4,000 by 25th, and up to 5,000 by the end of the month. Eight Common Gulls at the beginning of the week had increased to 300 by 25th, ending the month at over 500. A good 60 Great Black-backed Gulls present by the end of the month, too, but only 30 Herring Gulls.

There are still plenty of Lapwing and Snipe about, with a Curlew on 24th and 25th, and a Redshank on 30th; we see few Redshank here in the winter. The Short-eared Owls like the flood as it pushes the small mammals onto the banks and makes for easy feeding. We counted 12 feeding on Barrier Bank on 27th between the railway and Welches Dam.

There has been a Magpie about most of the week, although we see few of them as a rule. Long-tailed Tits have been very active again with daily sightings, 20 counted on 23rd and 30 the next day. Bearded Tits have arrived, too, one only at the beginning of the week, building up to a small flock of four by the end. These birds are most cooperative, spending their time in the Osier bed right in front of one of the new hides. On the morning of 30th while we were counting the Bewick's Swans there was, of all things, a Flamingo in the middle of the flock, quite unexpected and incongruous.

The early morning of 29th was misty and the Ouse Washes seemed out of this world. Millions of *Linyphiidae* spiders had been forced off the washes by the rising flood with the result that everywhere above water was covered in gossamer—emergent vegetation, fence posts, gates and Osiers—all shrouded in glistening thread. The beady gossamer, the greyness of the mist, and the mirror-still water, made an unforgettable scene.

DECEMBER

1st week. The first week of the month began with clear mild weather, the flood water rising to 178 cm (6 ft 10 in) on 2nd before slowly receding to 154 cm (5 ft) by 7th, but for the last four days of the week there was thick mist and we could only hear the birds.

On 1st December the water was creeping across the washes at a steady average rate of three inches a minute, stranding the small mammals and insects that live there in the dry spells. The voles and shrews can be found hiding in tussocks of Tufted Hair Grass which give only a temporary refuge before the rising waters push them off and they must swim the gauntlet of gulls, Kestrels and Short-eared Owls that are constantly working the flood edge. Many thousands of animals are caught.

By taking the punt up one of the ditches between two washes we were able to collect 48 Short-tailed Voles, six Common Shrews, two Water Shrews and a Bank Vole in half an hour of searching the tussocks. We had the little creatures running about under the duck boarding in the bottom of the punt until our return to Barrier Bank, when we released them. Doubtless, few would survive, displaced from their normal range and out of territory, but they had a better chance than before and it was a wonderful opportunity to get to know these mammals better.

The next day Gareth came down from Sandy, before dawn, and whilst it was still dark we took the punt to one of the Summer hides and spent the whole day there, recording the activity of the various ducks as part of his research work. Every half hour we would take a sample number of ducks and record the exact activity of each one. Was it feeding or loafing; how was it feeding; by upending in deep water or grazing away from the water, etc.? We were lucky and had an undisturbed day apart from two birdwatchers who stood on top of the bank—if only they had known how much they interrupted the feeding rate of the various species. I am sure that if some birdwatchers had occasion to see the harm they can do by thoughtless behaviour they would be much more careful not to cause unnecessary disturbance. I suppose it is difficult to realise how very different the wild duck on the Washes are from the ducks on the town lake.

Within view from the hide we had 150 Mallard, 50 rising to 100 Teal, and 3,000 Wigeon, but only 15 Pintail, five Shoveler and six Gadwall. This gave us useful figures on the Wigeon, Teal and Mallard but our other samples were a bit too low. The biggest surprise of the day was a Chiffchaff feeding in the half-submerged Osiers in front of the hide. We recorded four Great Tits, 12 Blue Tits, and 25 Long-tailed Tits passing through the Osiers during the day. As we punted home at dusk there were 11 Short-eared Owls working the bank between the hide and Welches Dam.

2nd week. The water level has dropped steadily during the week from 146 cm to 88 cm (4 ft 9 in to 2 ft 10 in) and, except where we had dammed the bottom of our washes to retain water in the refuge area, this left only shallow puddles at the lower ends of each wash. As a result, the duck converged on the wettest areas, and with the steady intake from the Continent, produced rising numbers on the refuge. Mallard numbers rose to 700, Teal 500; Wigeon are now over 6,000 strong on the RSPB refuge and even Pintail are up to 150, though the water is still a shade shallow for good numbers. Pochard dropped from 450 to only two birds during the week, the deserters retiring to the surrounding gravel and clay pits to await the next flood. A flock of 24 feral Greylag is still on the refuge, ignoring the din from nearby wildfowlers. Thirteen White-fronted Geese spent the day with us on 10th; if only the whole refuge could be secure I am sure we would get a regular flock but the present wildfowling pressure is too evenly distributed to give them a chance.

The Bewick's Swans increased steadily through the week from 160 to 281, 23 of them immature. It seems to be a less productive breeding season for them than last year, but it is early days yet. A Bewick's Swan hit the electricity wires which cross the Ouse Washes at Welches Dam on 13th and was killed.

At least the stomach provided useful information. Deaths from the wires are not as bad as three years ago when there was a lethal line of wires going over the washes at Welney. The Electricity Board was most helpful and removed the wires as they found they could reroute the current—there's hope for a future world yet!

A cormorant paid us a visit on 10th, perching on top of one of the hides and sunning itself most of the day. The hides are certainly used by birds; the viewing slots make favourite sheltered perches for the Kestrels and a good supply of regurgitated pellets can be found on opening any of the slots during the winter months.

3rd week. We made our last wildfowl count of the year on 17th. James, Brian, Gareth and I—a satisfactory day, good visibility, dry with a cold south-westerly wind. Not much flooding but the bottoms of the washes were very wet; the Delph depth only 64 cm (2 ft 1 in). The swans are building up well with 265 Mute Swans, 56 of them immature, 16 Whooper Swans, and 489 Bewick's Swans, 52 of them immature. There were still 22 Greylag and seven White-fronted Geese at Purls Bridge.

The duck numbers were excellent with 2,900 Mallard, 5,300 Teal (the largest number we have ever recorded on the Washes), 940 Pintail, 310 Shoveler, but only 18 Gadwall. The Wigeon count, at 19,500, shows that they are increasing satisfactorily and this must be our best-ever December figure. With 13,500 on the Welney refuge and only 440 in the shooting areas, the effect of shooting on the distribution of the wildfowl can be seen. As usual diving duck numbers were low; 120 Tufted Duck, 22 Pochard and a Golden-eye; 330 Coot and 230 Moorhen.

There were four Little Grebes on the Delph and Gareth had a Water Rail in the Old House Wood. The mildish weather for the time of year produced good numbers of waders: 38 Dunlin, one Curlew, 330 Golden Plover, 2,520 Lapwing and 16 Redshank, a high number for the time of year. In addition, 63 Ruff and 1,870 Snipe counted, but I wonder what is the true number of Snipe? They are hard enough to find on the ground at the best of times.

Quite high numbers of thrushes were recorded during the count: 830 Fieldfares, 5 Mistle Thrushes, 70 Blackbirds, 60 Redwings and 70 Song Thrushes. A flock of 200 Greenfinches up at Mepal; these are getting more numerous since I first came to the Ouse Washes. James had a Lesser Spotted Woodpecker, a Rock/Water Pipit and a Grey Wagtail between Mepal and Earith. We counted 13 Short-eared Owls and six Kestrels during the day; 66 species in all being recorded. Very few cattle on the washes now: 80 bullocks, plus 30 on the high banks, 30 horses, 60 sheep and a pig up near Mepal.

4th week. The year draws to a close with a slight flood. There was steady rain on 22nd and by 24th the depth had reached 115 cm (3 ft 9 in) and finished the month at 105 cm (3 ft 4 in). Very mild for the time of the year.

The shallow flood on the washes makes for excellent viewing of birds at Purls Bridge, and returning from Christmas fare on 27th we went to the hides on Barrier Bank which are well placed for conditions of this type. In the foreground the Bank slopes steeply downwards to the Osier Beds; Teasel and Thistles cover the slope and flocks of a dozen or so Goldfinches, with a score of Linnets and the odd Twite, were working through the plants. An occasional cheeky Meadow Pipit, Reed Bunting or perky Stonechat peered into the hide from a nearby Teazle head. The shallow flood waters covered the ground where the Osiers grow, and a Little Grebe swam up between the Osier bed and the bank, dipped its head under the shallow water to search for fish, diving from time to time to surface with the odd small morsel, some unfortunate stickleback, perhaps.

Beyond, in the river, which flowed sluggishly and discoloured, a group of six Tufted Ducks were diving at intervals; occasionally one came up with a large freshwater mussel, which took minutes to swallow. The damp, muddy banks of the river, well away from the flood, were covered with Snipe, and the flood edge on Barrier Bank and the washes were lined thick with Snipe. I have never seen so many; we counted well over 800 on the ground from the hide and there was a constant coming and going; flocks of Snipe in bunches of 40, 60 and 90 flying, almost like Dunlin, in tight weaving turns. There must have been several thousands on the Washes as a whole, and no doubt the mild weather had delayed their moving off to the south and west.

Not a large number of duck on the refuge, they have disturbed Christmases with all the shooting, but there was a good variety in front of the hides. Two or three thousand Wigeon on the nearest washes, most of them grazing behind the flood edge in tight packs of 200 or 300 birds. A good 1,000 Teal to be seen, mostly dabbling in shallows at the edge of the flood, and plenty of Mallard loafing about in the water with a few Wigeon and 50 or so Pintail. Occasionally a group of Mallard left to fly out to the Fens or came wheeling in off the Fens for a rest, one or two Pintail sometimes with them. These birds doubtless feed on the old potatoes left after the harvest and softened by the first frosts, so that the duck are able to take them. Twenty Shoveler, mostly drakes, fed in the water between the Teal and loafing Mallard, using their specially adapted beaks to sift the animal and insect foods from the water.

There were few swans, 12 Mute, a family party of six Whooper, five of them immature (I wonder what happened to the other adult?), 110 Bewick's Swans and only three of them youngsters. Many must have moved out over

Christmas, no doubt to quieter areas of the Washes. Eighteen Greylag seemed to have survived the holiday, and as long as they stay put they will be safe.

The longer we stayed in the hide the more it became apparent what we had previously missed. Twenty-four Dunlin feeding at the water's edge, so small they are easily lost among the tussocks. A good 200 Lapwing sprinkled about, mostly in the dryer areas. We counted 38 Ruff, distinguished on the ground from other waders by their solid, humpbacked, tatty-mantled appearance. In flight they are similar to Golden Plover with longer, thinner and more slowly moving wings, the body a bit Concorde-like with white oval patches at each side of the tail and very slight whitish wingbar confirming the 'jizz'.

We noticed a Hooded Crow together with nine Carrion Crows working the washes, searching for some unfortunate animal or injured bird not fully able to take care of itself. Considerable numbers of Fieldfares in the washes, obviously an influx over Christmas, not before time! As the afternoon drew on the light took on that yellowish quality, the far horizon hazing as the gulls started to pour in—at least 14,000 Black-headed Gulls, 200 Common Gulls, 300 Herring Gulls and 40 Great Black-backed Gulls. Most of them settled down on the water but some filled all the available space on one of the ditch-side spits that finger out into the flood.

We saw only four Short-eared Owls working the banks but had some superb views as they passed within a few feet of the hide, searching the bank for small mammals. It was quite a scene; quite a sound, too, with the noise from the fog of gulls, the honking Bewick's Swans, Wigeon whistling together and all the other natural sounds, undisturbed, the nearest main road a good five miles away. Such are our winter evenings, and it was with regret that we closed the shutters and walked homewards.

FARMLAND

by MIKE J. WAREING

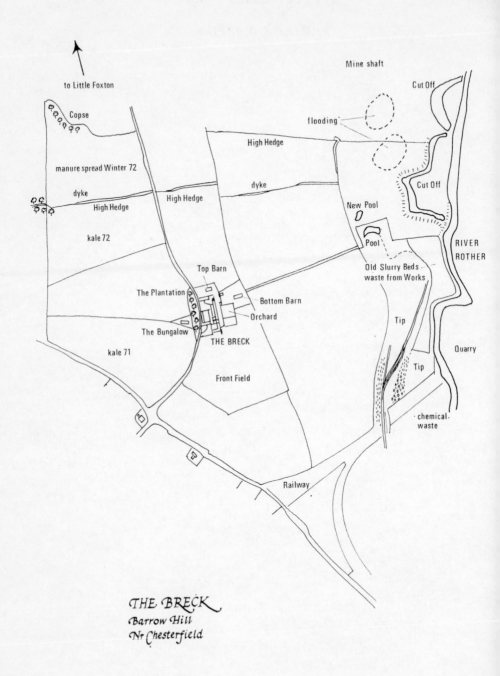

to Little Foxton

Copse

Mine shaft

Cut Off

flooding

manure spread Winter 72

High Hedge

dyke

dyke

High Hedge

High Hedge

Cut Off

kale 72

New Pool

Pool

Cut Off

RIVER
ROTHER

Top Barn

The Plantation

Bottom Barn

Old Slurry Beds
waste from Works

Orchard

The Bungalow

THE BRECK

Tip

Quarry

kale 71

Front Field

Tip

chemical
waste

Railway

THE BRECK
Barrow Hill
Nr Chesterfield

FARMLAND

9   The Breck—fields and farm buildings, from the tip.

10   The tip and railway beside the River Rother. Chemical waste, drums and carboys can be seen.

FARMLAND

11   Trees and scrub before open-cast working began.

12   The open-cast workings. The dark diagonal is coal.

Introduction

Before entering into an account of the bird life of a particular area it is necessary to discuss the type of habitat prevailing and the geographic, economic and social background that has led to its present development.

The Breck lies between Sheffield and Chesterfield in the Rother valley, and is surrounded by industry and rapidly developing urban communities. The main industries are coal-mining and chemical production, each of which have a profound, but not wholly bad, effect on the environment.

The land lies between the 100 and 300 ft contours, and is about 70 miles from the coast. It also lies in the foothills of the Peak District. These two geographical factors are important in several ways. It means that the Rother has a fall of about 18 inches per mile from the Breck to the coast, while in the other direction it rises sharply towards its source, so in the area round the Breck it shows a greater variation in level than any other river in England. Some of the Breck land lies below the flood level of the river, so drainage is difficult and the riverside pastures flood frequently with surface water which is comparatively unpolluted. This provides a good feeding area for ducks and waders in winter. The Rother itself is one of the most polluted rivers in the country but, even so, because it runs into the Don and from there into the Humber estuary it is a natural flight path for migratory birds, a number of which stay and winter with us.

The soil type is typical of those associated with coal measures. The basic structure is folded, with layers of sandstone, clay, coal, silt clay and glacial clay, covered with about nine inches of heavy loam. Where the different layers outcrop a great variation in the texture of the loam is seen. Where the glacial clay is near the surface it shows evidence of pressure ridges (see diagram).

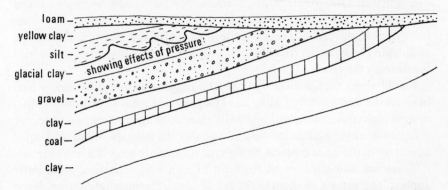

The glacial clay is practically impervious, resulting in water being trapped between the pressure ridges in the clay, and this leads to spring lines. These already difficult drainage conditions are further aggravated by mining subsidence damaging the existing drains. This situation makes for many difficulties from an agricultural point of view, but means that even in a period of drought there are small pools of unpolluted water available in most fields. These pools also have the additional advantage for bird and animal life in that they freeze only in the severest conditions. Where they occur in pasture fields the wet areas become 'poached' by the cattle and provide suitable feeding grounds for a wide variety of birds.

The historical background to an area plays a particularly large part in understanding its ecology. This is difficult to prove, but as far as birds are concerned our ringing programme, census work and observations suggest a striking correlation. The place names surrounding an area can often help by giving substantiating evidence in tracing the historical background. In the case of the Breck (Saxon, from brecklands, meaning wetlands) the neighbouring farm is called the Hagge, Saxon for a high place in a bog. The Parish Church is Norman, and in a neighbouring village there is a farm called the Grange, which derives from the Saxon for a barn or granary. These names all suggest that the area was used for agriculture as far back as Saxon times. Another interesting note is that the Hagge was rebuilt in the 15th–16th centuries as a hunting lodge. Proof of this exists in that there are still the old dog gates at the bottom of the stairs, and there is also a high balcony from which the ladies could watch the hunt. The fact that it was used as a hunting lodge testifies to the presence of abundant wildlife in the area.

These historical facts can be used to verify the age of existing hedgerows. Hooper suggests that the age of a hedgerow can be estimated by adding a hundred years for every tree or shrub species in it. Using his criteria, the oldest hedge would appear to date back to about A.D. 1000, having nine species in it. They are Oak, Maple, Elder, Blackthorn, Cherry, Ash, Sycamore, Hazel and Willow. This hedge has the additional ecological advantage of a dyke running alongside. Also, it is one of the lowest parts of the farm, and therefore sheltered. These factors contribute to make it the main arterial route for wild life through the farm.

Round about the farm there are several important ecological sites, which bear out the statement by Grime and Hodgson that man-made habitats often deserve high conservation ratings locally since some are now the most effective refuges for native plants. I would add to this the animals which take equal advantage of these sites, which, though not necessarily beautiful, are effective.

There are four sites round the farm to which this criterion particularly applies. The nearest of these to the house is also the most interesting, being 90% man-made and providing a wide variety of habitats. It is a large tip, consisting mainly of furnace slag with some chemical waste (which is partly being removed for road making), together with a railway embankment, the river Rother and an old sandstone quarry on the opposite side of the river which is now overgrown with Turkey Oak, Sycamore, Ash, Willow, Hawthorn, Alder and Elder, these being the predominant tree species. On the opposite side of the railway to the quarry lie about two acres of marshy ground, overgrown with nettles, thistles, willow herbs, Ragwort and Balsam. Areas of the slag tip provide a good example of how plants gradually recolonise a very poor and barren habitat. The most important species involved in this recolonisation are Rosebay Willowherb, Mullein, Thistles and Coltsfoot. The railway embankment used to be a mass of Meadow Cranesbill, but this died when the tipping started, and has now been replaced with various coarse grasses and Blackberry briars. The most amazing part of the habitat is the portion of the tip which is made up of a mass of old chemical containers, both plastic and metal, and chemical waste. This apparently apalling area provides a home for Rabbits, the occasional Fox and a wide variety of birds.

The next area of ecological interest is a piece of very rough land which was added to the farm when a new channel was cut for the river. This land is mainly spoil from the new river channel, and was left in a very uneven state and has gradually been recolonised by natural processes, and kept in check by grazing. There are some patches of Broom, a number of small Hawthorn bushes, and dense patches of thistles. The whole of this area is nearly surrounded by two ox-bow lakes formed when the banks of the new river channel were made. The old course was sealed off but was not filled in. These ox-bows are now free of pollution and have been stocked with fish by Staveley Works Fishing Club. The adjoining fields are rough grazing, regularly flooded by surface water.

Adjoining this wash land is an old mine shaft that has become overgrown with dense Hawthorn, Willow, Oak, Ash and Sycamore. This is very dense and provides a good roost for many birds in winter.

The fourth man-made habitat is about three-quarters of a mile from the farm and consists of a small lake, which serves as an emergency water supply for one of the local collieries. This is surrounded by about 100 acres of mixed woodland covering old ironstone workings. On the southwestern edge of this there is an area of scrub which lies on very rough ground, created first by ironstone mining and then by the removal of outcropping coal seams by miners in the strike of 1926. This five acres of scrub and Gorse (Little Foxton), is an extremely rich habitat both for wintering and breeding birds.

These four areas provide the main reservoirs for the wildlife on the farm, and the hedge-rows which link them are of great importance, emphasising the need for naturalists to consider their conservation. This linear conservation which I define as the conserving of a long narrow strip of countryside which links areas of ecological importance. Conservation in this way has many built-in advantages especially where it applies to old established hedges which are species-rich botanically. Hedges provide a large surface area, with varying conditions of light and shade, a wide variety of soil conditions, with moisture contents from very wet to dry and varying conditions of fertility. Consequently they are often the best refuge for a large number of plants which have been lost in our intensively managed pastures and arable land.

This variety of plant life in turn provides food and shelter for many forms of animal life. The large edge factor and the variety of plants that can be protected in this way make linear conservation an attractive proposition, and similar benefits are not obtained by conserving blocks of land with a small surface area. Additional advantages are that hedgerows are protected from public access by the surrounding agricultural land, which also provides a large feeding area for birds and small animals of the hedgerow.

The factors leading to the destruction of such a large percentage of this habitat are purely economic, and the only way to reduce the destruction is to provide countering financial incentives. The following tables give some idea of the costs involved, although since their compilation land prices have more than doubled and other costs have increased dramatically.

COMPARATIVE DATA ON HEDGEROWS

National Nature Reserves. 250,000 acres
Forestry commission survey shows there is approximately

> 52,000 acres of hedge in Devonshire
> 11,000 acres of hedge in Cumberland
> 13,000 acres of hedge in Derbyshire

Total farmable land in Derbyshire 474,622 acres.

TABLE OF COSTS RELATING TO HEDGES

All costs based on 1 square mile of land (640 acres)

> Cost of maintenance of hedges about £50 per mile per year
> Divided into 10 acre fields represents 16 miles of hedge to the square mile
> Divided into 40 acre fields represents 8 miles of hedge to the square mile

Where there are 16 miles of hedge to the square mile a hedge 3 yds wide represents 17.5 acres to the square mile

Taking Rent at £10 per acre

(i) 10 acre fields/16 miles of hedge per square mile/maintenance £800
(ii) 40 acre fields/8 miles of hedge per square mile/maintenance £400
(i) Rent £175/Total costs £975/Saving £488
(ii) Rent £87/Total costs £487

Taking Profit at £10 per acre

> Profit on 8.7 acres = £87
> Profit and Saving = £575

Loss of Agricultural Land

Average over past 10 years 57,000 acres per year.
This means an average loss of 1,460 miles of hedgrow per year or 1,575 acres per year without considering the loss due to removal and fire.

As can be seen from the costs shown, no farmer is going to be able to justify *to his accountant* the maintaining of superfluous hedges, whatever the cost in ecological terms. We have to accept that we live in a materialistic society, and that no other section of the community would accept this kind of deduction from earnings. The costs alone show the reasons for the disappearing hedge-rows, but there are other hidden factors which contribute, from the availability of labour in rural communities to the present tenant-right agreement, where the outgoing tenant is charged dilapidations on all uncut hedges. The only way in which this can be countered is by grants for laying hedges of twelve years growth or older, and charges in the tenant-right agreement such that only a percentage of the hedges should be trimmed and the rest in varying stages of growth. Where a hedge is going to be removed, then Ministry permission should have to be obtained, and the County Naturalist's Trust be given an opportunity to assess its ecological importance. If it is of value, then the farmer should have to demonstrate his need to remove it. If the reasons are valid then the Naturalist Trust, or the Department of the Environment, should make a contribution towards the cost of maintaining it, and in return be able to demand conditions on access and management. Management in this context would involve details of how and when the hedge was to be cut. This may be vital to conservation, for a Hawthorn hedge only fruits on wood which is at least two years old. Thus a hedge which is trimmed every year has far less to offer to wildlife, in that it has fewer berries for wintering birds and less in the way of nesting sites for breeding species. This means that a cycle has to be maintained in a hedge system, whereby there are always sections of the hedge at different stages of growth. These sections would then provide more food for insects in spring and summer and for birds during the winter months, and a habitat for nesting species which require higher or more overgrown nesting sites. The ideal situation for conservation is obtained when there is a system where one section of the hedge is trimmed, another growing up, and a third which is cut back and laid in order to recreate the cycle. The actual time of year at which a hedge is cut is also important. If a hedge is cut in autumn the food source and shelter for wintering birds is lost, and if it is cut too late in spring it disturbs nesting species. The ideal time appears to be early March, when the weather is beginning to pick up, food supplies are on the increase, and nesting species have not taken up territory.

Another aspect of management to be controlled is the type of spray used, and the time of year when it would be applied on and adjacent to the hedge concerned. This, coupled with grants for planting new hedges, should help redress the adverse trends of the past, and maintain these vital arteries through the countryside.

In the following text I hope to show the significance of my remarks, and how cropping and farm management can help wildlife on the farm whilst still maintaining an intensive system of agriculture. This will be done by considering a monthly record of bird life, compiled over the past four to five years, in conjunction with the farming routines.

JANUARY

When I started to look through the Ringing Group logs for January, I had hard work to convince myself that all the members of Breck Ringing Group had not hibernated during January and February for the last four years! The entries for these two months were virtually non-existent. Not until I looked at the ringing totals did I realise how completely ringing had dominated our activities, and how little birdwatching we had done. A total of 490 birds were ringed in January 1972, of which 293 were Greenfinches, and 127 were Bramblings.

The Kale for 1971–72 was grown in the field adjoining the bungalow. If we had deliberately set out to create a birdwatchers' dream, we could not have done better. Throughout December, January and February you could look out of the Bungalow window, and see at least a thousand birds without moving, and when, towards the end of the month, there was a week of cold weather, we were able to watch Brambling feeding on the bird table about six feet away. At one time there were at least 120 Brambling within 15 feet of the living room window feeding on 'tailings' (small grain and weed seeds) we had put down for them.

One Sunday after lunch we were watching about 50 Brambling feeding with the Greenfinches, Great Tits, Blue Tits, Dunnocks and Robins which were regulars at the bird table, when one of the children went to the other window, which looks out on the other side of the Bungalow, and called us to come and look. There were a dozen Goldfinches feeding on some chickweed not more than five yards away. Living in a rural environment has its advantages.

10th January 1973. Twenty White Fronted Geese flew over the farm in a northerly direction. There was the usual flock of birds on the Kale field, and six Redpolls were seen in the orchards.

In January and February 1969, as soon as the really cold weather started, we were able to watch an interesting duel most mornings and evenings. January 1969 the Barn Owl had been roosting regularly in the Bottom Barn, for about two months. During the first week of January the weather had gradually become colder, and we were getting some quite severe frosts at night. One clear frosty evening, as the light was beginning to fail, we were working in the Stack Yard, when we heard a commotion in the Bottom Barn and went to investigate. We were just in time to see a Kestrel swoop out of the Barn, and settle in the Ash tree near-by. After a few minutes the Kestrel made another excursion into the barn. There was a considerable amount of calling and hissing, and again the Kestrel retreated to the Ash tree. This happened another two or three times until the Barn Owl left on its nightly rounds. Later in the evening I took the torch, went down to the barn to look round, and there, roosting on one of the beams, sat the Kestrel. Next morning just before dawn I went down to the barn to see what would happen. I was soon rewarded. The Owl returned to reclaim its roosting place, swooped up into the barn, settled on the beam near the Kestrel, and proceded to hiss at and threaten it. At first the Kestrel was sleepy, and did not respond, until the Owl actually physically attacked. The Kestrel then reacted with such vigour that the Owl was glad to retreat to the Ash tree, and wait until the Kestrel was ready to leave. This skirmishing happened nearly every morning and evening until the end of February. The bird in possession was never driven out: it was always the incoming bird that had to wait. The only days when the battle was not joined was when the moon was very bright, and the Kestrel did a little nocturnal hunting, and came to roost long after the Barn Owl had left. We were most surprised when the same situation developed again in January and February of 1970.

A feature of winter that interests me greatly is the way in cold weather that the shyer species lose their fear, and come right into the cattle yard to feed. It is often possible to see up to 1,000 birds feeding in amongst the cattle. In extreme weather if you walk among the cattle you can get within a few feet of Cornbuntings, Skylarks, Meadow Pipits, Yellowhammers, Reed Buntings, Redpolls, Linnets, Chaffinches, Goldfinches, Bullfinches, Wrens and Dunnocks. This ability to adapt in extreme conditions must give a clue to the factors that have affected the dramatic increase in species like Starling, Magpie, Collared Dove and Blackbird.

FEBRUARY

I ended January by commenting on the way some species adapt, and lose their fear of man, particularly in extreme conditions. In February 1968 we had a classic example of this. There was a very cold spell of about three weeks duration. About 2,000 birds were feeding on the slurry we were spreading. One Pied Wagtail decided the competition in the field was too much for it, and found a method of getting in first. When the spreader was coming up the lane, about half a mile from the field, the Wagtail would fly to meet it, and sit on it until the field was reached, feeding all the way. It would stay on until the spreader was nearly empty then fly down and join the other birds. The Wagtail must have learned to recognise the sound of the tractor, for it always met the spreader at approximately the same place.

January and February on the Breck are very similar months as far as bird life is concerned. If cold weather blows up, all the Lapwings and most of the Duck move out almost overnight except for a few very hardy Mallard and Teal which attempt to survive on the Rother. How any form of life *can* survive after dabbling in the Rother defeats me! Farmers who have shot Mallard that have been feeding on the Rother say that they are black inside, and their flesh is so badly tainted that it is inedible. (The point of shooting them eludes me if they are inedible.) It would be very interesting if we could devise a way of catching some of these birds to ring them. We would then have a chance of finding out how long they survive after feeding on polluted water, and what their breeding success (or otherwise) is in the following years.

Mostly in January and February there is some mild weather and plenty of floodwater. Such periods are interesting for the large number of ducks that come in, and it is not unusual for there to be flocks of two or three hundred birds. Two typical days follow.

6th January 1968. Two shovelers, 120 Teal, 50 Mallard, six Pochard, 10 Tufted Duck, four Wigeon, 200 Black Headed Gulls.

3rd February 1969. Two Shelduck, 15 Tufted Duck, 10 Pochard, 80 Mallard, 140 Teal and three Bewick Swans.

In January 1972 a female Pintail was seen on the floods, and the first Wigeon I saw on the Breck was in the early part of February 1967. We had gone down on the Bottom Meadow to see that the main field drains had not been blocked. While we were there we went to the far end of the cut-off, and stood looking across the river at some flood water on a neighbour's field. We heard something rustle in a small clump of reeds growing on the edge of the water in the cut-off, only a few feet away, and on climbing down to investigate we found a Wigeon. It had been shot, and the tip of its wing was broken. The bird was otherwise undamaged but it would never have been able to fly again, so we took it to the Riber Nature Reserve, where its wing was treated and it was put on a pond with other wigeon. Guns!

One afternoon, early in February 1969, two young boys set fire to our nearest neighbour's barn, which was nearly full of straw. We went down to help, and had been there about half an hour when something shot out of the barn like a ball of fire, and hit the ground a few yards from us. It turned out to be a Barn Owl. I picked it up. It was lucky in that only the tips of its feathers had been burned: otherwise it was uninjured, but it would have to wait until it moulted before it would be able to fly again. We kept it for about a month, but the difficulty of supplying it with suitable food, persuaded us to take it to Riber Reserve where they have several other Barn Owls.

In February the Brambling were still feeding on the Kale stubble which had not yet been ploughed. With the help of a period of cold weather at the beginning of the month we managed to ring 207 Brambling and 188 Greenfinches. We caught them using clap nets as well as mist nets. From a flock we had originally estimated at 400 birds, we had ringed over 350 birds with less than 5% retraps, a clear indication that the real number was much higher. This alone shows how difficult it is to estimate flocks of small birds, and also how badly we had underestimated the size of the flock in the first place.

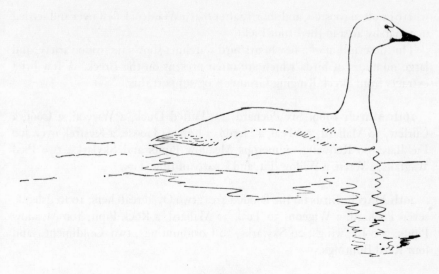

MARCH

March is one of the prime months in a diary of bird life. Resident species are taking up territory, the spring migration starts, and large numbers of wintering birds are still in evidence.

March 1969 was particularly interesting because of a prolonged period of extensive flooding on the Bottom Meadow, which attracted a large number of ducks and waders, some of which stayed on into the breeding season.

On the afternoon of the 14th March 1969, we went for a walk round the floods. There was little of interest on the permanent pool at the end of the tip, which, in spite of its highly polluted state, is usually good for waders during migration. We walked from there towards the cut-off and flushed three Grey Plovers—a first record for the Breck. As they rose, we heard a Rock Pipit calling near the tip, along with several Meadow Pipits. We continued towards the main body of flood water, and counted 70 Black-headed Gulls, 20 Lesser Blackbacked Gulls, 40 Mallard, and a pair of Shovelers. As we neared the hedge, a flock of 100+ Lapwings took to the air, taking with them 40 Teal and 29 Golden Plovers. Along the edge of the water we flushed a Snipe, and a Redshank. There were many Meadow Pipits and Skylarks about, but as these were scattered an estimation of numbers was difficult. The light was beginning to fail as we started to return to the farm following the High Hedge, from which we flushed many Tree Sparrows, Greenfinches, Blackbirds, Song Thrushes, Redwings and Fieldfares, and four Magpies. As we approached the farm buildings we saw a Barn Owl leave the Dutch Barn where he often roosted, and shortly after that a Woodcock flew over and settled in a marshy area in the Front Field.

This description of a few hours bird watching shows the wide variety, and large numbers of birds which are often present on the Breck. A few brief extracts from Breck Ringing Group's Log support this.

16th March 1969. Six Pochard, 12 Tufted Duck, a Wigeon, a Coot, a Curlew, 30 Mallard, 50 Teal, a Heron, a Canada Goose, a Kestrel, over 200 Fieldfare, 60 Redwing, numerous Meadow Pipits and Skylarks, two Pied Wagtails, a Wren, a Willow Tit, and a party of Blue Tits.

20th March. Birds on the flooded area: two Oystercatchers, 10 Redshank, seven Pochard, a Wigeon, 50 Teal, 30 Mallard, a Rock Pipit, 100 Meadow Pipits, 200 Lapwings, 20 Skylarks, 20 Cornbuntings, two Goldfinches and four Reed Buntings.

21st March is worth more detail. We were fencing some grassland adjoining the floods when, at about 11.00 hours, I heard a babbling noise, looked up and saw 65 Bewick Swans coming over in two large V formations. They circled and landed on the flood water. It was an unforgettable sight. The sun was shining, the sky clear, the blue reflecting in the flood water and making a wonderful backcloth for the graceful white swans settling on the water. The scene was made even more remarkable when contrasted with the grime of the chemical works, and the ugly urban sprawl of Staveley. The Swans stayed until 16.00 hours, then left in two flocks, one of about 45 the other 20. On the flood at the same time were 21 Curlews, four Redshanks, four Snipe, 10 Pochard, a Wigeon, three Canada Geese, 40 Teal, 50 Mallard, 1,000 Lapwings and the usual Skylarks, Meadow Pipits, Fieldfares and Redwings.

An extract from 30th March 1972 shows the similarities from year to year: two Snipe, two Wigeon, six Teal, 20 Mallard, a Coot, six Moorhens, 200 Fieldfares, 30 Redwings, 100 Black-headed Gulls, 50 Lesser Blackbacked Gulls, 400 Lapwings, a Wheatear, 20 Meadow Pipit, 30 Skylarks, 10 Linnets, two Pied Wagtails, two Corn Buntings and three Reed Buntings.

March is not one of our most successful months as far as ringing is concerned. The ringing totals graph for the month shows a spectacular drop from the winter, reflecting partly the break up of the large flocks of wintering finches but partly that less time was spent ringing.

March 1972 was remarkable for the unusually early start to the breeding season. On 1st March a Robin's nest was found in one of the nestboxes with one egg in it. A period of cold wet weather followed, and the nest was deserted.

On 4th March a Blackbird's nest with three eggs was found in the Covered Yard. This, as usual, was successful, and four young birds fledged. This particular nest is of great interest to me, for the same old cock bird uses this site every year, and is always an early starter. He has a white feather in his right wing, so it is easy to identify him. He is a remarkable example of the great adaptability of Blackbirds. His secret of success, even in a cold snowy March, is the way in which he robs the cats' food! I have seen this old bird come down to the cat food even when there have been as many as five cats round the dish, scold and chatter at them until they move back to a safe distance, and then slip up to the dish, grab two or three beakfulls of food, and dart back to his nest which is only about 20 yards away. He repeats this performance, completely oblivious not only of the cats, but also of people and dogs in the immediate vicinity. The nearness of his food source gives him an additional advantage in that he is always around to drive off predators threatening the nest, which he does with great energy, even the Magpies which manage to rob most of the other nests built in the farm buildings.

A Mistle Thrush, another aggressive bird, had a successful nest in a small apple tree in the orchard, but the other Blackbird and Song Thrush nests we found (even a Blackbird and a Song Thrush nest in one of the barns) were lost to either egg collectors or Magpies.

26th March 1972. We went for a walk along the river bank, behind the tip. This is an area of extreme industrial dereliction and pollution, typical of many that can be found in our industrial areas, but surprisingly, it is remarkable for the variety and numbers of birds to be seen there. In the last week of March, there are usually large numbers of Linnets returning from migration, as well as a few Goldfinches and Greenfinches. The day was proving to be much as I expected, so far as bird life was concerned. However as we rounded a bend in the river, we saw two men sitting on the bank about 300 yards away, and on looking through our binoculars we saw they had shotguns with them. Then a third man emerged from behind a clump of trees, and they made off up the railway embankment. The third man was carrying a Chardonneret trap, a type used to catch and cage song birds on the Continent. Pursuit was useless as they had too much of a start.

This incident, coupled with the surroundings in which it took place, highlights the pressures to which bird life is being subjected, not only from pollution and urban encroachment, but from poachers, egg collectors and bird trappers.

Bird trapping (for caging as distinct, of course, from ringing) is rife locally, and some men on the dole must be making a good tax-free living out of it. They have a ready market among some of the local breeders, and they have the time to find, and exploit, safe trapping sites. One well-known local dealer was prosecuted for having about 300 wild birds in his possession but the fine was no more than nominal. Bird trapping will no doubt continue, for a successful trapper can make up to £50 on a good day, and some illegally caught birds may go for 'export'.

APRIL

In April we see the major changeover from wintering birds to our breeding summer visitors. April 1971 produced some very interesting bird watching, and two new species for the farm. The afternoon of 6th was spent bird watching. It was a pleasant spring afternoon, so we set off down to the pool. Corn Buntings were in the hedge on both sides of the stackyard, and several Skylarks were marking territory by singing high over the surrounding fields. A Blackbird and a Mistle Thrush were seen carrying food, and as we approached the pool we heard a Little Ringed Plover calling (this is the earliest record we have). Round the pool we saw two Snipe, two Pied Wagtails, six Linnets and about 12 Tree Sparrows. In the long grass surrounding the pool two male Reed Buntings were singing against each other in a territorial dispute.

We went from there to the cut-off, where we disturbed a Heron taking advantage of Staveley Works Fishing Club's stocks. A small flock of Linnets was feeding in the rough grass on the opposite side of the water, and a Kingfisher flew past. As we approached the floods we flushed 15 Teal and 20 Mallard, both flocks predominantly males. We were on our way back along the High Hedge when, to our surprise, we saw a Buzzard gliding over in a north-westerly direction. This was the first record of a Buzzard for the farm. The High Hedge produced its usual list of interesting birds—three Robins, four Wrens, a Willow Tit, four Reed Buntings, three Yellowhammers, a small flock of Greenfinches, Tree Sparrows and House Sparrows, five Blackbirds, two Song Thrushes, 20 Fieldfares, two Magpies and three pairs of Dunnocks.

A few extracts from the Ringing Group log for April 1971 show the wide variety of birds seen on the Breck, and the times of arrival of the summer migrants.

12th April   A Kingfisher was seen on the cut-off.
14th April   Little Ringed Plover on the Pool. Willow Warbler in the
                High Hedge. Large flock of Fieldfares flying North.

15th April	Three Herons on the cut-off, and a Curlew on a pool of flood water.
17th April	Two Willow Warblers, a Jack Snipe and five Fieldfares.
18th April	The first Swallow was seen.
19th April	Five Yellow Wagtails.
22nd April	Five Swallows, two Yellow Wagtails, two Little Ringed Plovers, a Heron, a Whinchat, a Jack Snipe.
23rd April	Heavy rain for 24 hours—extensive flooding.
26th April	A Little Ringed Plover, a Redshank, a Heron, 13 Pochard, a male Shelduck, a Snipe, three Yellow Wagtails, a Wheatear, 25 Swallows, 16 Black-headed Gulls, two Lesser Black-backed Gulls and 20 Lapwings.
28th April	A Redshank, male and female Pochard, a male Shelduck, a Herring Gull, 39 Black-headed Gulls.
29th April	A Cuckoo, a Common Sandpiper, a male Shelduck.
30th April	A Whinchat, male and female Wheatear, a Common Sandpiper, 49 Black-headed Gulls, a Common Gull, 10 Mallard, two Teal and 10 Lapwings.

April 1969 cannot be passed over without a mention. Flood-water had accumulated earlier in the year, and persisted into May. Consequently there were large numbers of ducks and waders throughout the month. Four pairs of Mallard nested, and at the end of the month a pair of Shovelers were found to have nested, with two eggs by 30th. This was the first time Shovelers had been recorded breeding locally.

In April most of our resident breeders are firmly established. Several pairs of Starlings are very actively, and noisily, nesting in various holes in the stonework of the buildings. The Mistle Thrushes in the orchard and plantation are well away into the breeding season, and defending actively against allcomers, particularly Magpies and cats. The Blackbirds in the covered yard, and in the Holly hedge, are usually fledging their first broods by now, and fiercely attacking any marauders. Song Thrushes and Blackbirds nesting elsewhere are, as a rule, less successful as there is insufficient cover to protect them from egg collecting lads and Magpies.

This year we had a pair of Dunnocks feeding at the bird table until the middle of May. Naturally we became very interested in them and as time went on we developed the ability to recognise them as individuals. When courtship started it became easily possible to distinguish the male from the female by his actions, and to note other differences. The male was much slimmer, more grey round the head, and had brighter coverts and scapulars. The female was rounder and more Robin-like in shape, and her plumage was dingey compared with the male. The courtship of this pair was fascinating, the female displaying more actively than the male especially just prior to mating.

The Ringing Group's totals show that April is our worst month as far as ringing is concerned. As in March, this is mainly due to the dispersal of the large wintering flocks and also because less time is spent ringing. However April 1972 had its compensations in that we ringed our first Goldcrest—also the first recorded on the farm. Were it not for ringing, I am sure many of the less conspicuous species would probably go unrecorded. This Goldcrest was a classic example, as we had no idea that there were any about until we picked this bird out of the net.

April is the month when ornithologists begin to come into serious conflict with both legitimate shooting interests and poachers. Taking the latter first, the poachers, particularly those who are 'pot-hunters', shooting birds that have just paired or are already nesting, the result of their efforts can obviously be disastrous. The legitimate shooting interests are sometimes looking for something to keep their eye in, now that the game season is over. The most attractive targets are large flocks of Rooks, Feral Pigeons and Wood Pigeons, gathering on the newly sown spring cereal crops. This shooting is not in itself particularly harmful, but it is by no means as beneficial as the shooters would like one to believe. Shooting has little effect numerically on the pigeon population. All it does is move them from one feeding area to another for the period that shooting is taking place, and for a few hours afterwards. It also has some unexpected, and unwanted, side effects in that it may make it even easier for the pigeons that are going to breed (therefore the most important ones) to survive by making more food available to them.

MAY

It had been a bright, sunny spring day and in the cool of the evening the mist was beginning to form along the river, making attractive patterns in the hollows. We went down to the bottom meadow to see if there were any interesting migrants passing through. As we approached the first large body of flood water, we disturbed the Shovelers taking their evening feed. They sprang off, circled round, then went away into the mist towards the other flood-water, and their nesting site. A few minutes later the sound of a shot came from the direction of the nest, which was near the old mine shaft. We assumed that someone was shooting pigeons from the cover of the trees around the shaft, and so we took no notice. When we came to the end of the cut-off we saw the shoveler drake, lying injured in the water. We pulled him out, but he was so badly wounded that he died a few minutes later. We set off in the direction of the sound of the shot, when a man with a gun emerged from the mist. On being challenged, he denied having shot the Shoveler, saying he had permission from the neighbouring farmer to shoot Woodpigeon, Rabbits and Hares. He said he had been waiting for Hares (in May when they are out of season) when he had heard the shot, and he had come to investigate. We went across to the neighbour's farm and checked he had given the man permission to shoot. This confirmed, there was nothing more we could

210

do. The female Shoveler stayed around for several days, before departing, leaving her five eggs. So ended, sadly, the first recorded attempt of Shovelers to nest locally.

May 1971 highlighted another incident concerning guns. Some members of the syndicate which has the game shooting on the local farms had, throughout April and May, been keeping their eye in by very intensive pigeon shooting over the spring-sown crops. Every weekend there had been a morbid collection of wounded pigeons fluttering in the hedge bottoms and around the farm buildings. Towards the end of May we noticed some rather sick and emaciated feral pigeons sitting about on the roof of the barn. I decided that it would be kinder to kill them, as I thought they too were birds that had been shot, and wounded. I dispatched several, and to my surprise there were no shot marks on them. On opening their crops we found the cause of their illness. They had been picking up the small pellets of shot in mistake for charlock seeds. One bird had well over an ounce of lead in its crop! A very unusual way of poisoning pigeons.

May is always an exceptional month for the number of species seen. There is still a considerable movement of migrants, the breeding birds are settling to their territories, and a few small flocks of non-breeding birds are beginning to accumulate. May 1972 was one of the most interesting we have recorded.

6th May 1972. Two Swifts, a Little Ringed Plover, four Whinchats, a Blackcap, two Willow Warblers, a Grasshopper Warbler, two Swallows, and a Lesser Whitethroat which was caught and ringed. This was the first Lesser Whitethroat recorded on the Breck.

18th May. Two Turtle Doves, two Collared Doves.

20th May. We ringed our first Grasshopper Warbler.

29th May. Little Ringed Plover nest with four eggs found. A Carrion Crow's nest found, four Reed Bunting nests with four eggs. A Lesser Whitethroat, two Spotted Flycatchers, five pairs of Yellow Wagtails. A Pied Wagtail carrying food, two Collared Doves in the farmyard.

30th May. A Partridge's nest was found with 14 eggs, a young Little Owl seen, and a Red-legged Partridge was seen holding territory.

May 1971 was made interesting by extensive flooding.

1st May 1971. Two Redshank, two Little Ringed Plover, 13 Shelduck, 19 Wheatear, a Common Gull, a Sand Martin, 40 Black-headed Gulls, two

Lesser Black-backed Gulls, a Common Sandpiper, a pair of Whinchats, five pairs of Yellow Wagtails, a Swallow. Later in the day a Cuckoo was heard, a pair of Wheatears were seen in the stackyard, and the Common Sandpiper was caught and ringed.

27th May. An Oystercatcher was seen on the floods, a Curlew on the pool. Two pairs of Little Ringed Plovers were setting about nesting.

10th May 1970. A Little Ringed Plover, a grasshopper Warbler, a Tree Pipit, two Dunlin, 19 Wheatear, a 'Comic' Tern and a Curlew.

That month was also interesting in that Quail were heard on two occasions. Quail are uncommon in this part of Derbyshire, and there are very few breeding records. The diary for May would not be complete without the mention of three other unusual records for the area: a Bar-tailed Godwit was present from 3rd–5th May 1966, a Black-tailed Godwit on 15th and 16th May 1967, and two Sanderling on 16th May 1967. Strange to find so many unusual waders inland, all in this month.

May ringing totals are usually an improvement on the March and April figures, but the real interest lies in the variety of species caught and in the opportunity to study the changes that take place in birds at the beginning of the breeding season. Many species that cannot be sexed on plumage from midsummer to early spring show marked anatomical differences (such as the females developing brood patches) during the breeding season. Fortunately, this is particularly true of many of the warblers, the Lesser Whitethroat providing a good example. Work that we have done on Dunnock also shows that there are other features that could provide useful information. Tongue spots in the Dunnock are definitely a feature of breeding adults, and with further research could prove to be a method of ageing, and possibly sexing. Ringing in May has, on several occasions, proved the presence of some of the more secretive species that are often overlooked in the normal census work and general birdwatching. A recent example of this was the Lesser Whitethroat we caught on 6th May 1972. In the breeding season the ringer must be careful not to work near any nests, and must make sure that no bird spends more than a few minutes from being caught to being released. If this is done there is little or no disturbance to the birds.

During February, March and April the sowing of spring crops has taken place. The seed is usually dressed with one of the many insecticides or fungicides that are on the market. At this time of year the food for seed-eating birds is usually in very short supply, and any seed which is inadequately buried in the process of sowing is likely to be taken as food. This is the beginning of the 'chain effect' that has, for example, brought many of our predatory species near to disaster. Regrettably, this is the tip of the iceberg compared with the problems to be found in May. It is in May the limitations become apparent, in our own individual knowledge as farmers, in the lack of unbiased and well-informed advice, and in the overall gaps in collective knowledge of the side-effects of the chemicals now in use in agriculture.

In late April and early May cereal crops are 'top dressed' (i.e. the spreading of fertiliser on growing crops) and sprayed to control weeds, fungal attacks, and insect pests. Pastures are treated similarly. The use of these chemicals may justifiably be criticised in ecological terms. There is no doubt that the use of them may have a far reaching detrimental effect on the natural fauna and flora, but without their use it is doubtful whether even half our ever-growing population could be fed. The use of nitrogenous fertilisers is probably the greatest single factor in increasing our agricultural productivity. It is also becoming the major pollution factor arising from agriculture. Nitrates are highly soluble, and therefore leach out of the soil rapidly if adverse weather conditions prevent them being taken up by the plants within a short time of application, or if they are applied at the wrong stage of growth when the plants do not immediately require the nitrogen, and the nitrates are left in the soil at the mercy of the weather. This is a big problem from the farmer's point of view, for it means he is wasting money, but on the other hand it is not always practical to apply the fertiliser at the critical time. Thus a compromise has to be sought, which means the result is often far from the ideal.

Recently two other problems concerning nitrogenous fertilisers have come to light. First 'over application': this has occurred where intensive arable operations have been practised for many years, particularly where the mono-culture of barley has been practised. It has been found that the land has become so saturated with nitrogen that there is now very little response to it, and applying more nitrogenous fertiliser decreases yield. Fields such as this must be responsible for a large amount of the nitrogen draining into water-courses. Little is known of the effects that this saturation with nitrogen has on the natural fauna and flora of the soil, or to what extent it interferes with the uptake of other chemicals essential to plant growth. Second, the new problem created by the demand for high-protein grain. This will become of major importance now we have entered the European Common Market, for a pre-mium is paid on the protein content. A high protein content is achieved either by very late applications of nitrogen, or by the use of compounds that release nitrogen very slowly. Some of this nitrogen will inevitably be left in the soil to add to problems that are already severe. The only way that the point at which a particular field reaches saturation point can be found is by trial and error. These problems emphasise the need for more research and for advice which is freely given from an unbiased source. Otherwise we will make no impact on the problem.

The 'leaking' of nitrates is causing eutrophication of lakes, rivers and now even the sea, with far reaching effects on the ecological balance on a world scale. Some drinking water reservoirs, even, now have sufficiently high concentrations of nitrates to cause alarm.

Sprays are a similar problem: one which is made up partly of ignorance, partly of vested interest, and partly of competition. It is often regarded as a sign of bad husbandry if there are a few charlock flowers, docks or thistles showing in a field of cereals. Every year there are cases where farmers spend unnecessary money spraying to eliminate these, very often with a consequent loss of yield (and therefore profit) merely because they are not thoroughly acquainted with how much weed has to be present before it constitutes a threat to the crop, or what loss of yield is involved if the crop is sprayed after a certain stage of growth. Sadly, to this can be added fear of ridicule from their neighbours. There are cases where spraying has become an expensive yearly ritual, established partly because it is easier to spray than to have to decide whether weeds or other pests constitute a threat, and partly because out-of-season discounts encourage farmers to buy in advance before they have any real idea what the weed problem is going to be. The third and equally dangerous side to this problem is that of vested interest. The majority of technical advice on spraying comes from advisers employed by the chemical manufacturers. These firms have to sell to survive, and so do their staff. It is obvious that in a great number of cases their advice will be biased. These people are not going to miss a chance to sell, and they are certainly unlikely to advise the use of another firm's product, even if it is better for a specific case. To quote a case that happened in 1972, a farmer was advised that it was safe to spray grass with a selective weed killer while cattle were still using the field. Two animals died, and the toxicology department of the firm admitted that the symptoms were very similar to weedkiller poisoning, but liability could not be proved. This shows how imperative it is that greater control, and knowledge of the subject, should be in the hands of some independent body with no financial interest.

The recent reorganisation of A.D.A.S. means that it is now ineffective in too many cases, and I feel that an additional organisation, either connected with A.D.A.S. or the Department of the Environment should be set up. A policy that most of the technical advice to agriculture should come from the firms supplying agriculturists was one of the main themes of Mr Prior when he was Minister of Agriculture!—and the reorganisation of A.D.A.S. the start of the implementation of this!

JUNE

Again the question of shooting: I feel guns are too common in the countryside and too many people have access to them. A gun is a tool, not a toy to be used for pleasure or the personal satisfaction of some primitive urge. There are many excuses used for shooting, indeed some ornithologists suggest the shooting of one species to protect others. Magpies are an example of this. It has to be admitted there is sometimes a case for control, and that in these cases one of the most effective ways we have is shooting. I was going to add humane, but there are so few marksmen professional enough to kill without wounding a high percentage of birds and leaving them to die slowly, that such a remark would be inaccurate.

Magpies are becoming a pest largely because they are capable of exploiting an urban environment, living for a considerable proportion of their time on scraps obtained in gardens and around farm buildings. However, in the breeding season, they are very effective as predators, taking large numbers of eggs and nestlings. It is this that a number of ornithologists object to.

It was the shooting of a Carrion Crow on its nest that made me stop, think, and then start with these comments. The Carrion Crow, like the Magpie, is a very skilled taker of eggs and nestlings. Woodpigeon nests are comparatively easy to find, and so make up a considerable proportion of the Carrion Crows' diet. Magpies similarly are adept at robbing Woodpigeon nests. Of these three species the Woodpigeon is the pest of greatest economic significance agriculturally, and therefore nationally. It is also so prolific that it is not appreciably affected by shooting. In fact the Woodpigeon is so little affected by shooting, and is so economically significant, that experiments with various forms of

215

doped bait are being carried out. This is proving expensive, but will probably
be more effective.

It is just a thought, but might it be less expensive and more effective to let
nature find a balance? At the same time, however, we must remember that
we are all the time, by our ever-growing presence, let alone our manipulations
of the environment, creating a changing, and increasingly unnatural situation.
Research, and thought, is needed.

June is not notable for its rarities, but a few extracts from the log for June
1972 are interesting.

11th June 1972. The Little Ringed Plovers' nest was visited, and it was
found that two eggs had hatched and the other two were cold. One egg was
sent to Monks Wood Experimental Station for examination. This was broken
in transit, so they could not do a full pesticide residue analysis, but they did
say that embryo development had stopped at a very early stage.

17th June. A male Bullfinch was seen in the garden (the first seen round the
homestead in the breeding season). There were also six Wrens just fledged.

18th June. A Redshank, 32 Lapwings, 10 Linnets, two pairs of Whinchat,
five Goldfinches, a Whitethroat, a Lesser Whitethroat, four Yellow Wagtails,
and 10 Skylarks. The most interesting features of the day were the flocks of
non-breeding Lapwings and Skylarks.

23rd June. The morning was exciting in that it was the first time I have
been able to establish the presence of both male and female Quail on the farm.
Later on a Grasshopper Warbler was heard, and a Mallard and a Little Ringed
Plover were flushed from the pool. From this time on we never saw more than
one adult Little Ringed Plover, and saw nothing more of the nestlings. This
was the first time in four years that they appeared to have been unsuccessful,
and we wondered whether this was due to some poachers, who had been very
active in the vicinity of the nest.

25th June. Six Whinchat pulli (nestlings) were ringed. Later five Goldfinches, a Little Owl, two Whitethroats, a Willow Warbler, and a Pied Wagtail carrying food were seen. These were all at the back of the Tip. A well
known vet and ecologist of my acquaintance described this area as one of the
most horrific witches brews of pollution and desecration he had ever seen, yet
in 1972 it provided nesting sites for a pair of Little Owls, a pair of Little
Ringed Plovers, two pairs of Whinchats, two pairs of Pied Wagtails, 10 pairs
of Linnets, several pairs of Meadow Pipits, a pair of Whitethroats, two pairs

of Willow Warblers, three pairs of Yellowhammers, a pair of Robins, two pairs of Reed Buntings, a pair of Wrens and a pair of Common Partridges. My concern is to try and find out, in the long term, whether living and breeding in such a habitat affects the longevity of birds or their future ability to breed. Other users of this area are a pair of Foxes, which reared four cubs, many Rabbits, which can be seen playing in and out of the old chemical barrels, and at least one Weasel.

The next most popular breeding area is the High Hedge, and the bank of the dyke which runs alongside it. This supported, in 1972, a small colony of Linnets, two pairs of Whitethroats, a pair of Robins, two pairs of Wrens, several pairs of Dunnocks, Song Thrushes, Blackbirds, Reed Buntings and Yellow-hammers and a single pair each of Yellow Wagtails, Lesser Whitethroats, and Grasshopper Warblers.

Lying a close third to the High Hedge is the area round the homestead, which supports many pairs of House Sparrows, at least eight pairs of Tree-sparrows, five pairs of Starling, six pairs of Wrens, two pairs of Robins, four pairs Blue Tits, two pairs of Great Tits, a pair of Magpies, and two pairs of Swallows. The two pairs of Swallows in 1972 was the lowest I can remember, as our average is usually four or five pairs. As it turned out, 1972 was a disas-trous year at the Breck for Swallows, one pair laid and then deserted, rebuilt and were successful in hatching, but their young were taken. The other pair lost both broods in a similar fashion. How the young were taken we did not know. Magpies had been very active earlier in the year robbing Blackbird and Song Thrush nests actually in the buildings, and we assumed that they were probably the culprits.

The month cannot be passed over without a mention of June 1971, when the first pair of Coots to nest on the Breck successfully reared five young on the cut-off. A Moorhen nesting on the cut-off in the same year was not so lucky. One hot Sunday afternoon we took a walk down there and came to two fisher-men, who expressed concern that a Moorhen had not returned to her nest, in which there were four eggs just hatching. The young birds had emerged since they arrived, they told us, and both had died. Examination showed that the young in the other two eggs were also dead, and probably all were killed by exposure to the heat of the sun. The two men were fishing one on each side of the bush in which the nest was situated. They had both been within 10 yards of the nest since six o'clock in the morning. A straightforward lack of common sense, or selfish disregard?

In May and June, modern methods of intensive grazing and grass con-servation play havoc with the ground nesting species. This is something which is regrettably unavoidable if an economic return is to be made. The species that suffer the worst disturbance on the Breck are Corn Bunting,

Skylark, Partridge, Mallard, Yellow Wagtail and Pheasant. However, where there is also a large acreage of winter cereals these species have an alternative habitat, and are often forced to use this when their first nests are destroyed. The second nests seem to be sufficiently successful to keep numbers fairly static. In the case of Partridges and Pheasants, the second brood seems, even, to have a better survival rate. The young birds are, however, still very small at the beginning of the shooting season, to the dismay of the sporting fraternity. But this fact that they are small at the beginning of the winter, seems to have little effect on their ability to survive and be successful in following seasons.

Some individual birds seem to be much more resourceful and successful in the face of intensive grazing than others. One Skylark nest I found and watched provided one of the most remarkable incidents I have ever seen. The nest was in a very exposed tussock of grass, and it was impossible to protect it from the cattle, and when the cattle were allowed into the field I watched to see what happened. Whenever one of the grazing animals came within a foot or so, the Skylark came on to the edge of the nest and adopted a threatening posture, with her wings held out, and scolded vigorously. If this did not have the desired effect, she flew repeatedly at the cow's nose until it retreated! This amazing display of courage happily was not wasted, and the nest was successful. Later the same summer, I saw a Yellow Wagtail using similar techniques with equal success. These are the only two nests I have known survive this intensive grazing pressure, although there are no doubt others that have gone unseen.

JULY

It was a sunny summer morning early in July 1970, and the grass was still damp with dew. We were cutting some second crop grass for hay when we uncovered a Partridge nest with ten eggs in it. The nest was five or six yards from the hedge, and the eggs were so close to hatching that several of them were chipping. I debated whether to take the eggs to the house and hatch them under an infra-red light, but as Partridge chicks are notoriously difficult to rear, I decided against it. The old bird was still around and watching every move as I started to make a new nest in some of the uncut grass by the hedge-row, where it would not be disturbed by the mowing machine. Then I moved the eggs into it. Mowing was continuing, so I moved about a hundred yards away and watched. In less than a quarter of an hour the old bird had found the eggs, and settled down to brood them. I went back next day, by which time all the eggs had hatched, and the old bird had moved off with her brood. I had heard of this 'transplanting' being done before, but even so I was surprised when it was successful.

In July the Breck records show a considerable movement of bird life, as some unsuccessful and some non-breeders start to pass through. Young birds are plentiful. July 1968 was exceptional, the best we have in our records, when out of 210 birds ringed there were 60 Linnets, 41 Reed Buntings, 30 Tree Sparrows, 14 Yellow Wagtails and a Redstart. This adult Redstart was the first we had seen on the farm. The other birds were mostly juveniles. In July 1969 we were lucky enough to ring a Green Sandpiper which had been feeding on the Pool. July 1972 also had its rewards: we ringed a very young juvenile Lesser Whitethroat, and an adult with a very large brood patch. These were caught in the same net on consecutive days at the end of the month.

The bird log shows the movement of birds rather better than the ringing totals. A few extracts from 1972 illustrate the build up of Lapwing flocks in particular.

1st July 1972. Ringed a Whitethroat. Since the bad year for the Whitethroat population in 1969 we have taken particular notice of this species. 10.00 hrs: Grasshopper Warbler singing on the dyke side, four Goldfinches in the stackyard feeding on groundsel, 40 Swifts, 20 Sand Martins and four Yellow Wagtails round the pool. On the top fields were 70 Lapwings, 50 Starlings, and about 200 Rooks, one of which had white wings. The Blackbird in the Holly Hedge had three eggs, and this was her third brood. The first two broods had been successful, having reared four young each time.

2nd July. Seventy Lapwings on the top fields. 09.00 hrs. Two Rooks were seen attacking a Heron, which they forced to land on the cut-off. They then flew off.

3rd July. Forty Lapwings, a Golden Plover (a very unusual time of year for this species). There was a considerable movement of Tit parties and at 09.00 a Whitethroat was singing on the lane.

9th July. Eighty Lapwings.

21st July. Forty Lapwings at 06.00 hrs. A lovely sunny morning, and as I came back from fetching in the dairy herd, I saw a young Cuckoo on the garden fence. After I had watched and waited for a few minutes, a rather harrassed Dunnock came and fed it. Astonishing how such small birds can feed such a large youngster!

22nd July. 07.00. The young Cuckoo was again on the garden fence.

27th July. Two Turtle Doves were feeding in the Lane. Another juvenile Lesser Whitethroat was ringed.

28th July. An adult Lesser Whitethroat, with a large brood patch indicating that it was a female breeding locally, was ringed. Now a flock of 200 Lapwing flying over the top fields.

In July in years past I have noticed how large numbers of young birds, Linnets, Tree Sparrows, Greenfinches, Reed Buntings, some warblers and even Whinchats, tend to congregate around the fields where Potatoes and Kale are growing. This coincides with the time when aphid infestation of these crops would be coming to a peak, and spraying for aphids on the potato crops is likely to become necessary, especially where large acreages are grown. The amount of damage that the aphids are going to do is always difficult to assess. In a considerable number of cases, if the crop was *not* sprayed, I am sure that the cost of the damage would be less than the cost of spraying, especially if the Ladybird population is reasonably high. I have seen cases where crops with very high levels of aphid infestation have been left, and the ladybird population has very rapidly increased and controlled the attack, but this has always been in areas where insecticidal sprays have been rarely used. To attempt this policy in areas where large acreages of potatoes are grown, and insecticides regularly used for many years, would probably be to commit financial suicide, for there would almost certainly be insufficient predatory insects (e.g. ladybirds) left to make any impression on the aphid population. This leaves us with the roundabout that so often occurs where man interferes radically with the natural balance. We spray to kill the aphids, and in turn we kill the beneficial predatory insects, either by the direct effects of the insecticides, or by the removal of their food supply. Using these steamroller tactics, we also kill many insects that play a vital roll in the pollination of plants, and of course we destroy an important part of the food source of many young birds (shooting fraternity please note that young Partridges are among these); and we lessen the seed production which is so useful as the staple diet for wintering birds. These wintering birds would, in their turn, have helped in future years to keep insect pests under control, and damage down to an acceptable level. Some way has to be found that is economically acceptable, to break this cycle of ever-increasing interference with the natural balance. I certainly do not feel that more and more supposedly improved chemicals are the answer.

AUGUST

The Breck log for August is rather brief, for the obvious reason that this is a farm, and harvest is in full swing throughout the month, leaving little time for anything but work. The records that are available show that many birds have started to move back to their winter quarters, that the flocks of young birds continue to grow, and that the great majority of birds have finished breeding.

2nd August 1970. Two Curlews flying East, a Little Ringed Plover on the Pool.

3rd August. Three Snipe, and a Wheatear on the pool. A large flock of finches in the High Hedge, and 20 Swallows near the farm buildings.

22nd August. Two Kingfishers on the cut-off, a Sedge Warbler, a Blackcap and 10 Meadow Pipits at the back of the tip.

In August 1968 some interesting birds were ringed: two Yellow Wagtails, three Mistle Thrushes, a Redshank, a Whitethroat, a Snipe, three Common Sandpipers and 34 Tree Sparrows. The Tree Sparrows are just one indication of the build up of the flocks of young birds. In August 1969 we again ringed three Common Sandpipers. This is a species that only occurs on the Breck during migration.

The log for 1972 shows the continuance of the same trend.

3rd August. Eight Snipe, and a Green Sandpiper on the Pool. A Little Ringed Plover, 40 Linnets and four Collared Doves on the tip. Also seen were 130 Lapwings, 50 feral Pigeons, 40 Wood Pigeons, 100 Rooks and 26 Swallows.

7th **August.** A Curlew, 200 or more Lapwings, four Willow Warblers, and 150 Swallows.

August on an arable farm means hard work for long hours, but there are many compensations for those who find interest in wildlife, and who are lucky enough to be working in the open rather than shut in with a hot dusty grain-drier for company 16 or more hours a day. If the weather is good, it can be a time of excitement, beauty and interest. If the weather is bad it can also be time of depression, tension and desperately hard work.

There is no work on the farm that I enjoy more than combine harvesting on a fine summer evening. I well remember one such evening, late in August 1967, with the sun setting in a blaze of glory after a scorching hot day. We were cutting spring barley, and there were about 500 Swallows, Martins and Swifts sweeping back and forth over the field for insects disturbed by the combine. Several Reed Buntings were trying hard to find suitable roosting places in the small patch of barley that remained. As the combine approached a small patch of weather-flattened barley, I noticed what appeared to be a large brown stone in the middle of it. When I came nearer and was able to see it more clearly. I realised that the 'stone' was an owl of some kind. It was not until it took to the air that I was able to recognise it as a Long-eared Owl. It must have been just resting, for when it was disturbed it flew slowly away, and settled in the next field of Barley. (This, so far, is the only Long-eared Owl to have been seen on the Breck.) It is surprises like this that add a great amount of colour and interest to the work.

The previous day also had its moments of interest. Early in the morning a Corncrake had flown out of the barley in front of the combine (the last Corn-crake seen on the Breck: a few years previously they had been common). Later in the day two coveys of young Partridges were disturbed, and in the evening two adults and four young Pheasants flew out. Just as we were finishing the field, a lovely dog Fox ran out, his coat flashing red in the evening sun.

There are also many interesting plants in the stubble, especially in fields that have not been sprayed with herbicides. They are often only small, but some of them I find particularly beautiful, especially on close examination. Three of my favourites are Germander Speedwell, Scarlet Pimpernel and Field Pansy, and I would be very sorry to see those species wiped out. Some species that already seem to have been virtually lost to arable land locally are Cornflower and Common Red Poppy. I can only hope that we do not lose too much of interest and beauty to the increasing demands (both economic and for high output) made on agriculture today.

SEPTEMBER

On average, harvesting is finished on the Breck by the end of the second week in September, and cultivation of the stubble is in full swing. This means that there is a plentiful supply of food for all species of birds, and large flocks gather, sometimes so numerous that estimation of numbers becomes almost impossible. Most birdwatchers would find it a fascinating experience to be on a tractor, ploughing the stubble prior to the sowing of winter cereals, for most birds seem to be totally unafraid of tractors, and barely bother to hop out of the furrow to allow the tractor to pass. This gives opportunities for the most marvellous close-up views of many species like Wheatears, Pied and Yellow Wagtails, Song and Mistle Thrushes, Blackbirds, Robins, Dunnock and Linnets which are usually the most bold of the smaller birds. I have occasionally seen even Willow Warblers, Redstarts and Spotted Flycatchers, along with the Whinchats, Meadow Pipits and Skylarks, that you would expect to see feeding in this way. The larger birds commonly associated with this method of feeding are Lesser Blackbacked Gulls, Black-headed Gulls, Herring Gulls, Lapwings, Rooks and Jackdaws. These are so unafraid of tractors they often come within two or three feet, and it is possible to see whether or not they are ringed. Other species I have had good close-up views of are Golden Plover, Ringed Plover, Common Gull, and Greater Blackbacked Gull.

In most years, a feature of September is the large flocks of hirundines that gather prior to migration, and some years there are flocks of several hundred Swallows, one or two hundred House Martins, and similar numbers of Sand

Martins. This year, however, was very disappointing, the largest gathering of Swallows being 150 seen in August. House Martins and Sand Martins were down to groups of about 20. This trend was fortunately not consistent in this part of the country, for on one farm nearby, I saw between 200 and 300 Swallows on the 26th September, and was told that four young Swallows had fledged that day from a nest in the buildings.

Some extracts from the Breck Log for September will help complete the picture.

9th September 1972. Eight Turtle Doves, 15 Collared Doves, 80 Wood-pigeons, 200 Rooks and 15 Magpies feeding on stubble, 30 Lesser Black-backed Gulls 120 Black-headed Gulls, and 150 Lapwings on some newly ploughed ground. A Jay in the plantation, four Snipe and 12 Yellow Wagtails on the pool, 50 Swallows around the farm buildings.

17th September. Six Swallows around the farm buildings. A flock of finches in the High Hedge included at least 35 Greenfinches, four Gold-finches, two Corn Buntings, and 40 Tree Sparrows. On the Bottom Meadow were 57 Lapwings, a Golden Plover, 27 Snipe, 11 Mallard, a Teal, a Curlew, eight Reed Buntings, four Yellow Wagtails and a Kingfisher. Round the tip were 30 Meadow Pipits, and a Whinchat. In the other fields on stubble and ploughed land were 100 Rooks, 50 Black-headed Gulls, 30 Lesser Blackbacked Gulls, a Common Gull, two Turtle Doves, 10 Collared Doves, 10 Magpies, two Kestrels and a Little Owl. This list would be much longer if we had included species like House Sparrows, Dunnocks, Robins, Wrens, Blackbirds, Song Thrushes and Starlings which are everyday species on the Breck as elsewhere.

On the morning of 17th September, the children were feeding the few hens they keep as pets when they disturbed a Fox, which had apparently settled down for the day in a patch of nettles at the back of the barn only a few yards from the chicken shed! This was not the first time that a Fox had chosen to spend the day in the stackyard. There was an incident a few years ago, when we were keeping hens commercially on 'verandahs' in the stackyard. On the morning concerned, I had filled the water fountains for the hens, and was just starting to fill the feed hoppers, when I noticed that the hens were behaving rather curiously. They were all looking down through the slatted floor. I eventually looked underneath, and there curled up, quite comfortably and fast asleep, was a large Fox. It must have sensed my presence for it woke with a start, and made off over the orchard wall, and away down the fields. A close examination showed the Fox had been living there for several days. This incident was made more surprising when one considers that there are always four dogs in the stackyard, and that tractors and people are moving about all day. It is another demonstration of the ability of some animals to adapt to an urban environment.

20th September. A Green Sandpiper, nine Mallard, a Snipe, over 400 Linnets, 200 Greenfinches, 50 Tree Sparrows, 15 Swallows, 40 Lapwings, three Cornbuntings, five Meadow Pipits, six Skylarks, a couple of Reed Buntings and a Little Owl.

29th September. Two thousand Lapwings, 300 Linnets, 400 Green-finches, 200 Tree Sparrows, seven Swallows, four Moorhens, 30 Snipe, a Jack Snipe, a Yellow Wagtail and 30 Meadow Pipits. It is interesting to note how the Lapwings build up gradually through the autumn.

Before leaving September, the potato crop has be be considered, because early in the month it would have its final spraying with fungicide to prevent blight, and towards the end of the month it would be sprayed with Diquat to burn of haulm before harvesting. With the sprays for blight and haulm desiccation a warning is given to allow at least a week before harvesting and to keep farm animals away from the sprayed fields. But what price wildlife! How do we warn them? The other side of the coin is that without these sprays the potato crop would fail more times than it would succeed. Land usage is full of these dilemmas, none of them easily resolved.

OCTOBER

October is the month when winter cereals are sown. It is here I must state that as a farmer I am totally against the use of any very persistent insecticide, fungicide, or herbicide, whatever the cost in economic terms. Aldrin (traces of this chemical have been found in soil eight years after application) is still in limited use in agriculture (and practically unlimited use in the average garden!) with about 20% of the winter-sown wheat dressed with it. It is also still on the Ministry of Agriculture's recommended list for the control of Cabbage Root Fly, Hop Root Weevil, Leatherjackets (DDT-sensitive varieties of spring Barley) Narcissus Bulb Fly, Wireworm (Potato only) and as a winter seed dressing against Wheat Bulb Fly. A statement in the Ministry booklet on Approved Products for Farmers and Growers says 'Uses of the chemical have been limited by *agreement* under the Pesticides Safety Precautions Scheme to those shown above. This substance is a lasting, and indiscriminating killer of many forms of soil fauna (the beneficial along with the harmful) as well as having lasting and undesirable effects on many higher organisms.' The appalling thing to me is that most of the pests shown above can be controlled or kept within the limits that will still allow viable cropping, either by good management or by the use of insecticides that are short-term in their action. A similar comment can be made about Dieldrin. These two chemicals, and several others (DDT being by far the best known and most widely used) that fall into the same category, are still in World-wide use, and are so persistent that they are to be found throughout the oceans of the World, even into the Antarctic. I think that in permitting the use of these persistent chemicals, whose long-term effects on World ecology are as yet so utterly unpredictable, the human race is being profligate with its environment, and future generations will perhaps be faced with insoluble problems or incalculable losses, not to mention actual threats to survival. It is easy to argue the case for the undoubted benefits that chemicals like DDT offer mankind, especially in the tropics, but in view of the equally undoubted dangers, have the possibilities of alternatives really been exhaustively explored?

In October the Breck is at its peak for birdlife, having its highest numbers and greatest variety. Again a few extracts from the log to illustrate this.

14th October 1970. Three Snow Buntings (an adult male, a first-year male, and an adult female). These were on the tip at the edge of the pool.

15th October. Two Snow Buntings—the first-year male and the adult female.

16th October. Snow Buntings still present, eight Swallows.

18th October. A Redstart, by the pool.

Extracts from the log for 1971 also show the unexpected:

4th October 1971. One Oystercatcher, 30 Swallows and a Kingfisher.

12th October. Three juvenile Swallows on the wires over the Stack Yard. Two Jays in the plantation, a Jack Snipe and six Golden Plovers.

18th October. A Goldfinch, 100 Mallard, a Teal, a Jack Snipe, a late Swallow, 50 Redwing, over 100 Meadow Pipits.

31st October. A Goldcrest, 110 Redwings, 300 Fieldfares, 10 Goldfinches, 50 Linnets, 100 Lesser Blackbacked Gulls, Jays still actively feeding on the large crop of acorns on the Turkey Oaks.

The log for October 1972 perhaps demonstrates numbers, rather than the unexpected.

1st October 1972. A thousand Lapwings, a flock of over 1,000 finches made up of Linnets, Tree Sparrows, House Sparrows and Greenfinches. These were feeding in the kale. A Yellow Wagtail on the pool. Six hundred Rooks, 40 or so Meadow Pipits, 30 Reed Buntings, 20 Corn Buntings, four Moorhens.

8th October. Two Swallows, five Snipe, a Meadow Pipit, over 400 Linnets, eight Mistle Thrushes, 20 Corn Buntings, 80 Chaffinches, 400 Tree Sparrows,

200 House Sparrows, 500 Greenfinches, a Curlew, 100 Lesser Blackbacked Gulls, 200 Black-headed Gulls.

21st October. A hundred Golden Plover, the largest flock I have seen on the Breck. It was a clear sunny morning after rain, and the birds were settled quietly on a pasture which was lush and green in spite of the time of year. Really this was the first time I realised that Golden Plover are so aptly named. I had always thought of them as the dull brown silhouettes one sees as they pass over mixed in with a flock of Lapwings. There were also 2,000 Lapwings, 150 Black-headed Gulls, 1,000 Starlings, 400 Fieldfares, 50 Redwings, 700 Woodpigeons, 100 Linnets, 200 Tree Sparrows, 100 Greenfinches, 20 Meadow Pipits, 50 Skylarks, 200 Chaffinches and 20 Reed Buntings. In the rough grass round the pool we disturbed a Fox, and later we saw a Weasel hunting on the dyke side. Quite a day!

31st October. One thousand Lapwings, 1,500 finches in the Kale, and a very late House Martin flying round the farm buildings.

Ringing in October 1968 brought to light two species we had not suspected were present on the Breck. A Treecreeper and a Blackcap in the net! The first record for both species, and in fact there have been no further records of Treecreepers.

In October 1972 we were discussing the possibilities that the large finch flocks, which were feeding in the Kale, were roosting in Little Foxton. Our idea was to try to ring them there, the following weekend. One evening we decided to go and have a look to see what was roosting there. The number of roosting birds was not as large as we had hoped, but there was certainly a good variety: Coal Tits, Blue Tits, Great Tits, Willow Tits, Long-tailed Tits, Goldcrests, Chaffinches, Bullfinches, Greenfinches, Yellowhammers, Blackbirds, Song Thrushes, Tree Sparrows, Jays and a pair of Woodcock. We decided we would give netting a try the following weekend.

On the morning of the 15th, I saw my first Fieldfare of the winter, and this made us even more in favour of ringing in Little Foxton that afternoon, as in the winter it provides roosting for up to 1,000 Fieldfares and about 500 Redwing, and we were hoping that some would have arrived. Early in the afternoon we set off complete with poles and nets, and all the other trappings that go with ringing. There is a thick hedge surrounding Little Foxton, and from the direction we approach you can see nothing of the scrub until you get right up to the hedge. Imagine the surprised look on our faces, when we looked through a gap in the hedge, and saw nothing but brown earth! Five acres of scrub and Gorse, with a small stream running through it, had been completely flattened by the bulldozers from the adjoining opencast site! The owners of this opencast working had obviously decided to extend their operations! Five acres of extremely interesting and varied habitat, that had taken between 60 and 100 years to become established, was gone in less than a week. It is obvious man cannot go on eliminating blocks of countryside in this way, unless he makes some serious effort to recreate similar habitats elsewhere. This takes time, land and money—all precious commodities, but we *must* afford them.

NOVEMBER

All our summer birds have gone, the days are shortening, and by the end of the month all the cattle will be housed in the yard. This always signifies to me the start of winter, and the four most difficult months of the year, as far as my farming is concerned. When the cattle are housed there is always a significant increase in health problems. All the food for the cattle has to be carted to them, and keeping the yards clean becomes a full-time occupation. Cold, wet, and sludge dominate the picture. In short I have a strong natural aversion to winter!

Bird life, however, is still abundant and unusually concentrates in four main areas. The Kale field, the homestead, the floods and the field where we spread the slurry from the yards daily. This is shown in extracts from the log:

6th November 1972. Three Snipe in the Bottom Meadow and a Heron on the cut-off. The Kale field had its usual flock of finches, 20 Brambling, 50 Corn Buntings, three Goldfinches, 500 Linnets, 500 Greenfinches, 100 Chaffinch, six Reed Buntings, two Bullfinch and 200 Tree Sparrows. On the field where we were spreading slurry were 50 Fieldfare, 50 Redwings, 400 feral Pigeons (an astonishing number), 50 Woodpigeons, 10 Magpies and 500 Lapwings. In the copse, two Long-tailed Tits possibly displaced from Little Foxton.

14th November. There had been heavy rain, and there was extensive flooding on the bottom fields. On the floods were 20 Teal, 15 Mallard, 20

Snipe, a Jack Snipe and 500 Lapwings. Feeding on the slurry were 300 Rooks, 40 Jackdaws and about 1,000 Starlings. On the same field were a further 1,500 Lapwings. In the Kale were the usual large flock of finches and Buntings.

15th November was very cold and provided a contrast in that the Lapwings had moved out overnight. The floods had frozen and there were 30 Teal and 20 Mallard sitting in miserable huddles on the ice. The bird activity in the Kale was unchanged, but the feral Pigeons feeding in the slurry had been joined by about 50 Lesser Blackbacked Gulls and 70 Black-headed Gulls. In the evening it was interesting to see that a Barn Owl had started roosting in the straw bales in the top Barn.

16th November. I was going down the land to the Kale field, when I saw a Kestrel swoop into a flock of finches and take one of the larger birds (probably a Greenfinch). It had difficulty in holding it, and settled in the field about 100 yards away to adjust its grip. It had only just settled when it was attacked by three Magpies. The Kestrel had great difficulty in evading the attack while keeping a firm hold on its prey. It eventually escaped and sped away as rapidly as it could, hotly pursued by one of the Magpies, which followed it for at least three quarters of a mile before giving up. This was the first time I had seen Magpies attack a Kestrel in this way. The 16th was a bad day for the Kestrel population, for later on I saw a Kestrel harassed by two Rooks for about a quarter of an hour before it eluded them!

19th November. With the mild weather, the Lapwings had returned, and there was about 2,000 in three flocks scattered round the farm. There was little or no wind, and the mild humid weather soon made us acutely aware of our most undesirable neighbour, the Chemical Works. The air was so heavy with fumes that it made breathing difficult and uncomfortable. What the acid content was I hate to think, but a blue haze hung over the valley all day, and the next day the grass on the lower fields was spotted almost as if it had suffered from the drift from spraying with weedkillers. We are luckier as a rule, because the prevailing wind takes the fumes away from us, carrying them directly over Staveley. Surely fumes released in this kind of concentration cannot benefit the health of the local population, and must have very damaging direct effects on the environment.

28th November. During the last few days I have been noticing massive movements of Starlings. A flock of over 5,000 birds has passed over every morning at about 8 a.m., breaking up into smaller groups and dispersing as it travelled away in a westerly direction. The flock gradually gathered in the afternoon for a final feed on the field where we are disposing of the slurry, and at about 4 p.m. left, in one mass of birds, in an easterly direction. We often have very large flocks of Starlings at this time of the year, but this is the largest concentration I have ever seen locally, and judging by the other flocks I have seen I would think that there is a roost of about 30,000 birds not very far away.

In the afternoon of the 28th I went down on the Bottom Meadow to see that none of the drains had been blocked by debris carried down by the floods following the heavy rain on the 14th. Most of the floodwater had subsided, and there were about 20 Teal and six Mallard circling round, and as I walked across the Bottom Meadow I flushed about 20 Snipe. I then noticed something hanging in the top of one of the hawthorns in the High Hedge, and as I came nearer I realised that it was a dead bird. I climbed the bush (about 15 feet) at considerable expense to various parts of my anatomy to discover that it was the corpse of a Moorhen, whose feet had become entangled in a length of fishing line. The line had subsequently become firmly entangled round a branch, imprisoning the bird, and it had died either of exhaustion or starvation. I have also found a Blackbird and a Yellow Wagtail which have suffered similar fates, in the same hedge. I only wish that the fishermen of the world would take heed, and not leave lengths of fishing line about. I also wish that they would take their bottles and other litter with them. The foot of a cow which has stepped on a bottle is not a pretty sight, and it does not help the economy of the farm when a cow dies from eating a polythene bag which had previously contained someone's lunch.

November cannot be left without mentioning an unexpected sight that occurred in 1967. On 12th November 1967 a large bird was seen on the floods. A member of the Ringing Group went down to investigate and was amazed to see a Cormorant sitting there. It left the next day having, we assumed, found that there were no fish present!

DECEMBER

In December, Kale proves not only its value to the farm economy, but also its value to conservation. Twelve to fifteen acres of Kale are grown every year on the Breck. This is 'zero grazed' (meaning that it is cut with a forage harvester and fed to the cattle in the yards fresh every day), and lasts from the middle of November until the middle of January. During this time it provides a roosting place, and a feeding ground, for literally hundreds of birds, sometimes as many as three to four thousand. The species that use it as a roost are Wren, Dunnock, Linnet, Greenfinch, Reed Bunting and Yellowhammer. It also serves as a feeding ground for these species, and for Chaffinch, Brambling, Corn Bunting, Pied Wagtail, Song Thrush, Blackbird, Fieldfare, Lapwing, Golden Plover, Woodpigeon and feral Pigeon besides. Naturally with these large numbers of small birds about, the predators take advantage, so Kestrel, Little Owl and Barn Owl are active. Kale also provides cover for game birds like Partridge, Pheasant, Snipe and I have also seen Woodcock. In winter, cover is scarce, it is a refuge for a number of animals: Foxes, Stoats, Weasels, Hares and Rabbits. The one disadvantage is that it has all been harvested by the time the winter is at its worst, and this means that all these species that have benefitted from it have to disperse to find food and shelter elsewhere.

The Breck Ringing Group log shows that December 1972 was a particularly interesting month. There was torrential rain at the beginning, and by the 8th there was extensive flooding. On the bottom Meadow were 50 Teal, 20 Mallard, 30 Black-headed Gulls, 50 Fieldfare, 40 Redwing, two Snipe, a Golden Plover and 1,000 Lapwings. On the slurry field were a further 500 Lapwings, 400 feral Pigeons and 6,000 Starlings. The flock of 'small birds' were feeding as usual in the Kale field, and there was a Jay in the Plantation.

In the afternoon I saw another unusual incident concerning a Kestrel which was hunting over the hedgerow by the Bungalow when it was mobbed by about 20 Starlings until it was driven for shelter into one of the Ash trees on the Lane. The Starlings were still flying round the tree when three Magpies flew into the tree and chased the Kestrel across the Front Field, where it finally managed to escape.

9th December 1972. The children had been playing round the floodwater, and when they came home they said they had seen a Water Rail. I found it difficult to believe that they were not mistaken, as it was at least fifteen years since I last saw one on the Breck. However I thought I should check, and had a walk down next day, and was lucky enough to flush the Water Rail from the side of a small bush by the High Hedge.

13th December. From the buildings we saw some large white birds on the floodwater: six Bewick's Swans. On the bottom Meadow at the same time were 100 Lesser Blackbacked Gulls, 15 Great Blackbacked Gulls, 50 Black-headed Gulls, 40 Teal, two Carrion Crows and 500 Lapwings. The Bewick's stayed until the 15th.

17th December. Most of the floodwater had gone, 10 Bewick's were seen flying over and two Shovelers, nine Teal and 20 Mallards were disturbed from the small pool of floodwater that remained. The flock of small birds in the Kale had increased to over 3,000, and with them were 600 Lapwings. The Barn Owl was still using the top barn.

19th December. A Kestrel again featured in an unusual incident: I had just opened the door to one of the calf boxes, to take in some bedding, when I saw an escaped yellow Budgerigar, sitting and looking very miserable on the covered-yard fence. The gap between the building and the covered-yard is about 12 feet, and the tractor was parked there. I quickly shut the door, with every intention of trying to catch the Budgerigar. I turned, and carefully started to approach, when there was a loud swish, an explosion of yellow feathers, and a male Kestrel lifted up over the buildings, clutching the unfortunate bird in his talons. I was not more than six feet from the Buderigar when the Kestrel struck.

December 1971 saw the build up of the largest flock of Bramblings I have ever seen. By the end of the month there were an estimated 400 birds. But it was not until the end of February 1972, that we realised how bad we were at estimating large numbers of small birds!

RINGING

Some birdwatchers may fear that ringing causes disturbance. Thus I feel that it is imperative that the value of ringing for various projects should be shown. Without ringing it would be utterly impossible for us to find out anything about the movement of birds to and from the Breck. It would be impossible for us to study population changes. It would also be very difficult for us to know what effect on birdlife, any changes in our farming policy had. Just watching the birds is too imprecise, for we would be unable to distinguish one individual from another. To illustrate this I am including the records of some individuals that I find particularly interesting.

From our recoveries of Willow Tits we find that they are the Breck's most sedentary species as we regularly retrap the same individuals in the same place. Their territory is very limited, though in severe conditions they will change slightly. They are also very persistent in their use of their favourite flight path as the records for individuals caught regularly in the same netting site show.

	Date Ringed/Retrapped	Time	Winglength mm	Weight gm
JJ87382	10.08.72	09.00	56	11
	10.08.72	10.00		11
	10.08.72	14.00		11
	06.11.72	15.00		10.5
	05.01.73	10.00		11.5
HP87652	22.09.68	13.00	59	10.0
	25.09.69	19.00		11.5
JJ54727	22.01.72	15.00	57	10
	05.02.72	13.00		9.5
	08.07.72	08.00		10
	26.07.72	09.00		10
	29.12.72	12.00		10

We have had no recoveries off the farm, and it could be inferred from this information that any destruction of habitat (such as the removal of hedgerows) could be very damaging as far as this species is concerned.

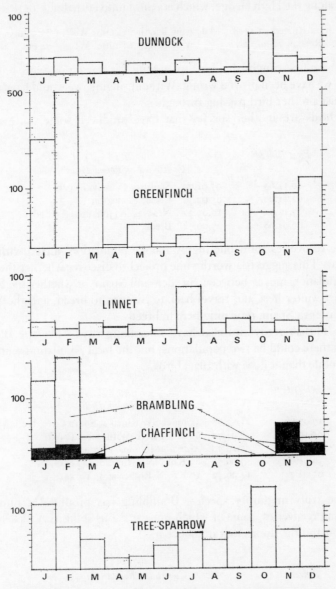

Some of the more common species ringed at The Breck, showing
numbers ringed 1969–72

This year we had a very interesting retrap: a Great Spotted Woodpecker, feeding along the High Hedge, which is a most unlikely habitat for this species.

First Ringed 22.11.70 Male wing length 132 mm Weight 77.5 gm
Retrapped 29.12.72   wing length 133 mm Weight 74 gm

When we first caught this bird we assumed it was on passage, but subsequently we have been proved wrong. Without ringing, we would have thought it was just another bird passing through.

Blackbirds are another species that have produced some quite startling results.

First Ringed	Killed Controlled/or Found Dead*	
16.11.68	07.01.70	Santurce (Vizcaya) Spain
02.11.69	05.02.72	Local, Eckington
07.11.71	10.05.72	Nr. Nassjo (Jonkoping) Sweden
07.04.68	27.09.72	Breck

These recoveries show one highly mobile population, mixed with a very static one. This suggests a worthwhile project to discover whether the migratory population moves between Sweden and Spain, or whether the Swedish birds just winter here and travel back to Sweden to breed, and the Spanish birds winter in Spain, returning here to breed.

Song Thrush recoveries also have interesting twist—like the Blackbird it seems there could be two populations, but the local population seems to be more mobile than it does with Blackbirds.

Ringed	Killed Controlled/Found Dead	
17.07.66	21.12.68	45 miles SW Rined at South Cave, Hull (Yorks)
05.10.69	17.04.70	63 miles ENE Withernsea (Yorks)
02.11.69	15.10.70	Marcillac (Gironde) France
30.11.69	17.06.71	63 Km NNW Bradford (Yorks)
05.02.72	14.06.72	17 Km N Rotherham (Yorks)

Of the truly migratory species, Brambling has produced a number of interesting recoveries, some of which must make us think very closely about migration and the things that influence it.

Ringed	Killed Controlled/Found Dead	
16.02.69	24.07.70	Kolpisjorvi, Entekio (Lappe) Finland
30.01.72	18.10.72	Trier, Rhienland-pfalz (Germany)
67	30.01.72	Breck

* 'Retrap' is the term used to describe a bird caught again by the original ringer; if it is caught by another ringer, it is called a 'control'.

This last bird was ringed with an avicultural ring when caught and the dates on the ring suggested that the bird was five years old. With the number of local trappers catching birds illegally, it was possible that it did not give the bird's correct age.

A Redwing we ringed 25.03.71 was later shot at Lenola, Fondi (Latina) Italy 06.01.72. What were the migration routes of this Redwing and the Bramblings? Why did they winter in different parts of Europe in consecutive winters?

Linnets on the Breck also show interesting movements. It seems from our recoveries that the birds which breed on the Breck usually winter in France and Spain, but as yet we know little of the origins of our wintering population.

Ringed	Controlled/caught caged	
05.05.68	20.10.68	Novafria (Segovia) Spain
08.05.69	21.10.70	Biscurrosse (Landes) France

Greenfinch is the most common species on the Breck. Last year about 700 were ringed. It is surprising that flocks of over 2,000 birds can gather from such a small area. Greenfinches, although very mobile in a given area, do no seem to move very far outside it. All our recoveries fall within a 75 km radius, the predominant movement being north-westerly. We have found that few of the birds that winter on the Breck breed there, and we are hoping to obtain more information from the large number of birds ringed in the last 12 months.*

Tree Sparrow is another species that winters at the Breck in large numbers: it is surprising that we have had only one recovery that was not local. This bird, like most of the Greenfinches, had travelled 50 km in a north-westerly direction. This north-westerly movement is very puzzling.

Although we ring very few waders at the Breck, we have been very fortunate with recoveries:

Redshank ringed 18.05.68   found dead Hallen, Nr Bristol 05.01.70
Snipe ringed 09.08.69   controlled Trewalder, Camelford (Cornwall) 07.04.71

Ringing has proved that most of our breeding Dunnocks are sedentary, but in October 1968 there seemed to be something of a population explosion. We were almost certain some of the birds ringed in this period were of continental origin, but unfortunately we have had no recoveries to prove it.

Blue Tits and Great Tits are other species in which the breeding population is sedentary, and we are now recovering birds that are at least four years old. However there is a more mobile population that winters on the Breck.

* Since this text was drafted, a Greenfinch ringed at the Breck has been controlled in Aldershot, Hants, 230 km SSE. Greenfinch ringed 24.02.72 controlled Aldershot 31.12.72.

As can be seen from these few notes, considerable information is being obtained from our ringing programme, information that could be gained in no other way. It is this sort of precise knowledge that is required if we are going to attempt to conserve many of our species in this age of economic pressures, to protect them for our ever expanding urban population, so that they are there for future generations to enjoy and make use of.

CONCLUSION

In this diary I have tried to show the variety of our local bird life, and the interest that it can engender. Our birds, and the countryside of which they are but a part, are something that, as a generation, we ought to pay more attention to conserving. Man is destructive and greedy by nature. He demands more and more from his environment as his technology advances, and as the population increases. One thing that agriculture teaches is that you cannot go on taking without making the effort to replace whatever it is that you have taken. An old saying is 'starve the land and it will starve you, feed it and it will feed you': the same is true of our environment.

Every species we destroy is an irreplaceable loss to the Earth's genetic pool, as well as a loss in itself, and we never know when we will have need of it. Examples of the sort of use we can make of this pool are:

In the last few years the value of the Musk Ox has been realised, and a commercial herd of these animals (which were hunted to near-extinction) has been set up in Scandinavia, and it is hoped that herds will eventually be set up through the whole of northern Europe including the highlands of Scotland. They can live on very sparse vegetation, they provide meat, and it has been found that their wool is exceptionally fine and has very good insulating properties (so good, in fact, that it is used in undergarments for astronauts).

Bramblings, amongst others, are being studied to see how birds can rapidly put down large amounts of fat prior to migration. Some of the results of this work may eventually help us to gain some understanding of obesity in man.

Plants with a high tolerance of lead can be used to recolonise the spoil heaps from lead mining, which are otherwise clean of vegetation and most unsightly.

We can never predict what demands we will need to make on the Earth's range of plant and animal species or genetic pool, but the larger the pool is, the better chance we have of keeping our ecology in balance. This does not mean that I feel that a species should be preserved just in case it can be commercially exploited at some future date, because obviously there is far more to a healthy environment than that.

Crop spraying: open any farming journal, and you will see pages devoted to advertising one or other of the many big chemical company's products. These quotations were taken at random from one farming journal.

'Why Risk It? Fasten on the May & Baker "Safety-Belt" Weedkillers.'

'Fylene for Black Grass control? Great! Even better than the Fisons chap predicted.'

'The Big Plus for Cereal Growers—Bayer Yield Builders.'

'Di-Farmon. There are over 50 common British weeds which can take up residence in your cereal crop. They'll take up the nutrient that you put down for the crop—and profit too.'

'Di-Farmon is a broad-spectrum herbicide which controls over 40 weeds in cereals.'

These high-pressure salesmen are the people to whom the Government has suggested that we turn for advice on fertilisers and sprays. Can anyone really see the Fisons chap telling a farmer to buy Bayer's 'Big Plus' or a May & Baker rep saying 'Di-Farmon' would be best?! It surpasses comprehension that anyone could be so naïve, particularly a Minister of Agriculture, but such was one Minister's suggestion! These firms cannot run on altruism, and how can the average farmer cut through the jungle of extravagant claims to reach an assessment which is the most effective material for his individual needs? There surely needs to be some organisation, with no financial interest, to assess farming needs, offer advice, and above all to control use, paying attention to environmental needs as well as to economic advantage. Think of the ecological damage a herbicide with the range of 'Di-Farmon' could do. Forty species of 'weeds' alone—how many attractive, and perhaps important food plants for other animals or birds will be eliminated from our countryside in such a sledge-hammer operation?

There is also the economic situation that practically makes the destruction of our hedge-rows a prerequisite, where an all-arable policy is pursued. Straw burning, to which so many people object (and rightly), is carried out simply because it is the easiest and most economic way to deal with the surplus.

We live in a materialistic society, and farmers, like businessmen, cannot survive on altruism. This may sound like an apology for modern farming techniques, but surely the demands for a high quality product, free from insect or fungal blemish, stem from an affluent consumer, and the demands for ever-higher yields are directly related to the apparently thoughtless rate at which the population of consumers is increasing itself.

As I sit writing this last paragraph I can look out of the window, and see Greenfinches, Blue Tits, Robins and Dunnocks feeding at the bird table, further away I can see Lapwing, Golden Plover, and Black-headed Gulls circling in the bright sunlight, against a background of the chemical works, and a blue haze of fumes pouring out over Staveley. The Breck struggles for survival as farmland on the fringe of a greedily expanding industrial area. The farmer has long faced the threats of land-gobbling by houses and factories, and now has the problems of the rapidly increasing commercialism of farming itself to counter. It must make us all think, and wonder what kind of future we are creating.

This diary is dedicated to Sallie, Quentin, Fiona and Mark, and I am pleased to acknowledge my gratitude to G. P. Mawson, M. Stott, G. Burman and R. A. Frost.

MOUNTAIN AND MOORLAND

by DONALD WATSON

The Dark Gairies

JANUARY

In this diary of a year's birdwatching in a region of mountains, moor and forest in South Scotland I have chosen five sample areas well known to me, referred to as the Moor, the Flow, the High Glen, the Dark Gairies and the Seven-mile Ridge. In recent years afforestation by coniferous trees has affected the character and bird populations of all these areas, to a varying degree. This hill country is never very far from the lowlands, with rivers, lochs and marshes set in farmland noted for its herds of dairy cattle and winter wildfowl. The proximity of this very different landscape brings contrast and charm to the region and often influences the movements and variety of birds found in the hills. All the sample areas conform to real localities, but the names used will not be found on the Ordnance Survey maps.

1st January. New Year's Day, the sun shining in an ice-blue sky; crisp snow on the moundy fields above the village. The outline of the high western hills has a fragile beauty under snow. Once of Himalayan proportions, the Seven-mile Ridge still has a commanding profile, never dropping below two thousand feet. Snow picks out the vertical rides in the dark skirt of forest on the slopes. I turned north up the High Glen; birds were sparse in these conditions. A pair of Ravens flew high overhead almost touching wings, their resonant and sometimes liquid song notes, far-carrying in the clear still air. This may be a tree-nesting pair which builds in a tall Scots pine in a shelter wood on the open moor. They will roost near the old nest tonight. At the deep cleuch at the head of the glen there is another pair which always nest on a rock-site. How far do Ravens travel from their bases in search of food? The other night I watched a pair at dusk, flying in a perfectly straight line until beyond vision with binoculars; the site they were presumably heading for was a good eight miles away and they had come from further afield.

There was no Peregrine at the cleuch today but the herd says that the pair have been about lately. A month ago I watched one come in to roost at another eyrie. There is evidently little to attract predators up here just now—on the way I only saw a cock Kestrel making rather laboured progress from one tele-graph pole to the next—but my diary reminds me that on 4th January 1964 there were at least six Short-eared Owls, as many Kestrels, two Hen Harriers and a Buzzard hunting around a field of unharvested oats near the river. That was a very mild winter and no doubt rodent numbers were high in that field—perhaps even still breeding. Short-eared Owls are usually scarce in these hills, in January, even on planted ground. Four days later, in fact, the Owls had all gone, but I noted a single harrier and, on the hill near by, a fox stalking voles, characteristically standing up on its hind legs, tail in the air, then leaping to pounce into a clump of rushes. Down in the valley, recently, a keeper caught five foxes in snares in one night; wintry weather may have brought many down from the hills. Often obvious enough to the human eye these snares are deadly to foxes and are widely used, in the forests and outside them. Of course dogs and other animals are at risk too.

The snow is not deep enough to drive the Snow Buntings down. There are a few on the Ridge just now but the big flocks, which were commonplace even at stackyard level in the 1930s, are rare in this district now. The reasons for this change are not known. Reed Buntings are the most noticeable small birds near the farmhouse half way up the glen, feeding with a few Chaffinches and a Blue Tit around the cattle troughs. This glen is notable for the long straggle of willows which line the river bank—scarcely more than a burn here—and extend out onto the boggy floor of the valley to form patches of scrub wood-land. Even in mid-winter there are Willow Tits and a Great Spotted Wood-pecker here and a flock of Fieldfares seemed set to roost in the bushes—they are not always ground roosters. A Tawny Owl perched low on a branch at the waterside was perhaps looking for Water Voles which I have seen here. Unlike the three other owls of the district—Barn, Short-eared and Long-eared—the Tawny does not hunt by coursing the ground but almost always drops on its prey from a perch. Telegraph poles along moorland roadsides enable it to extend its hunting well away from woods. Long-eared Owls have been known to breed in a Crow's nest in willow scrub not far away but they do not seem to be here; in this weather they might be seen hunting well before dark. Barn Owls, of course, are out hunting by early afternoon on these cold days; if bad weather persists many will die.

Blackcock are the chief glory of the 'sauchs' as the willows are called here, but it is early enough in the year to see many feeding on the buds; a party of eight, with no Greyhens in attendance, were quietly feeding on seeds of grasses, none showing the least sign of a red comb. There are Pheasants up here too—two magnificent cocks, breasts burnished by the low sunlight, were high up in a hawthorn tree, eating the last crimson berries. Both Pheasants and Blackgame often mingle with the finches and buntings where the cattle are fed.

15th January. Jimmy, the herd at the Moor, has been seeing a lot of the Eagles. A pair have been about his inby fields for over a week. His wife was rather alarmed to find one sitting on the farm road when she went to open the gate. If I had not known this oddly domestic pair of Eagles I should have thought she had seen a Buzzard. They obviously know the shepherd and accept him as part of the landscape.

I walked up on to the little ridge, past the stell with the twisted old larch, once a memorable hide for watching a Hen Harrier's nest, and sat down in a crook of heather to enjoy a favourite landscape. At eight hundred feet the view is still spacious, with a foreground tapestry of heather, moorgrass, bracken and bog myrtle sweeping down to the boggy 'Fleshmarket' (a name given to bog or precipice alike, each treacherous to sheep). But the forest covers nearly all the further ground, beneath the high snowy tops, and has recently extended over the neighbouring hill farm. My favourite moorland has become almost an island in the ocean of conifers and may soon be submerged.

Back at the roadside I scanned the edge of a ten year old block of sitkas and lodgepole pines. A small grey and white blob became a Great Grey Shrike, flashing black and white as it swooped to the ground. This is a typical haunt of wintering shrikes; there are at least five different territories in forest of fairly similar age in the district. The same areas are re-occupied, sometimes, but not always in successive years, invariably by single birds, but it cannot be assumed that the same individual returns. The late Duchess of Bedford recorded one as long ago as 1907 at a spot where one has occurred in recent winters—at that site hawthorn trees, not conifers, provide most of the hunting perches. Looking under a thorn bush I recently found small pink and black feathers from a Long-tailed Tit killed by the shrike. In the forest, voles are important food as suggested by the behaviour of the bird today, frequently dropping into long grass at the forest boundary. Bird watchers often report Great Grey Shrikes—the combination of comparative rarity and conspicuousness is conducive—and sightings through a winter suggest that some individuals range more widely than indicated by published accounts of 'home ranges' of about one hundred acres. Raymond Hewson found 83% field vole and 17% bird prey, with traces of invertebrates in the pellets of a forest-haunting Great Grey Shrike on Speyside.

24th January. Mild rainy weather; a flock of nine Oystercatchers flew up the valley. Single birds occasionally come up the river in early January but the first flocks are a promise of spring.

An hour before sunset we are at the edge of the forest, looking out towards the Flow, the level boggy valley used by Hen Harriers for communal winter roosting. We are careful to keep in the cover of the trees, to minimise disturbance. The first arrival is a ringtail, excitingly close overhead; thirty-five minutes before sunset. The dark silhouette against the still light sky clearly shows a full crop on this bird. It turns and drifts on the wind, quickly down to grass-top height, hesitates once and drops with lowered feet among tall moorgrass and rushes, vanishing swiftly. For ten minutes there is no sound or movement except the wind and the rain; daylight fading, curtains of cold rain, desolation—why have we come? Then the blood quickens again when a grey male and a ringtail appear out of nowhere, and almost immediately another ringtail; three birds in the air at once, as if searching the ground, before dropping down for the night. At half an hour past sunset we think the total has reached twelve, but only two are old males. In four out of five years the January maximum of numbers using the roost has been very stable at twelve to fifteen but counts are always much lower than this in cold, frosty conditions. It is by no means certain that fewer individuals are in the district in cold spells. Why should the weather affect numbers at the roost? The many fascinating questions posed by observations at the harrier roost must await later discussion.

After the last harrier had settled the black shapes of Carrion Crows were still crossing from the open hill towards their forest roost, mostly in widely spaced couples.

At night, in the misty darkness, Golden Plover were calling as they flew over the village.

FEBRUARY

1st February. Quiet grey weather, colours as rich as fruit cake under a high cloudy canopy. Brilliant Goldfinches and Siskins on the riverside alders, a drake Goldeneye, subdued, luminous blue-white among dark reflections under the bank. Last night was the final spree for the inland duck and goose shooters—there will be a few wounded geese down the valley this morning.

Far up on the forested hill, at the back of the Flow, the 50 year old larches are well-thinned now and attract many small birds at this season; at first the trees seemed birdless but I was suddenly beneath a roving flock of nearly 100 Coal Tits and Goldcrests with a few Chaffinches. The forester had reported Crossbills and I had a good view of four, including two brick-red males, on the larch cones. There may be fledged young about as breeding in the neighbouring county was in full swing in November–December. Low temperatures in no way inhibited them but many broods died in persistent rain. H., who learned his forest skills in the mountains of Yugoslavia, is a master at tracking Crossbills to their nests, which are usually high enough to be difficult to spot from the ground, but in January I saw a local nest, beside a forest road, which was fairly obvious against the trunk of a sitka, at about 30 feet from the ground. A good cone crop also brings big flocks of Redpolls and Siskins to the tall larches, the Siskins often singing in chorus. Some years ago a self-styled 'vermin killer' brought me a beautiful cock Siskin dropped by a Sparrow Hawk he had shot at the edge of a larch wood.

A Buzzard rose, mewing, near a known nesting tree where I have seen a pair fly in to roost on a winter evening. An old photograph (c. 1920), of a nest with eggs on a rock face confirms that Buzzards bred here before the present forest came. A pair or two continued to find sites, usually in remnant hard-wood trees, until the larches and sitkas were big enough to build in. The forest is certainly the safest place for nesting but they hunt the open ground in and across the valley, where they are often shot or trapped by keepers. Persecution still keeps the numbers of Buzzards far below the capacity of this region. At first thought it seems remarkable how rarely we see Buzzards overflying the Flow when watching the harrier roost but evidently the surrounding moorland and young forest have little to offer them compared with the abundance of food, particularly rabbits, about the valley farms and woodland edges which lie in the opposite direction. The continued presence of Eagles on the hill ground may also be some deterrent to Buzzards which, like Peregrines, are certainly not permitted to share a nesting crag with a pair of Eagles. The flight-lines of the Hen Harriers, to and from the Flow, show that they too hunt more in the lowlands than on the hill-ground at this season.

7th February. Stopped to observe the Ravens at the Little Gairy. They were at work on the lower of their alternative sites, generally preferred as a first choice. It is less accessible than it looks, being well overhung, at the top of a rock fissure. I saw one bird fly in with a long stick in its enormous bill, then the pair at the nest together. It will be interesting to see how long a pair will continue to nest here, now that it is surrounded by young forest. In the last 20 years several sites have been deserted as trees have replaced sheep and this could be the next to go. Forestry policy of drastically reducing the 'wild' goat flocks further limits carrion supplies in these hills, and if Ravens go short of food one wonders about the future of the Eagles at Ben Gairy. Both Ravens and Eagles follow deer hunters, to feed on the grallochs, but the availability of these depends largely on the open seasons for stalking, when the annual culls of deer are made by the forest rangers. The seasons are worth recording, as few people carry the dates in their heads. They are, for Red deer; Stags 1st July–20th October, Hinds 21st October–15th January; and for Roe Deer; Bucks 30th April–21st October, Does 20th October–1st March. Thus there would only be the grallochs of Roe Does in January–February and none at all (except when deer are shot out of season because of extensive tree damage) in the crucial early spring period when carrion is probably most important to Eagles, Ravens and, sometimes, Buzzards.

Although close to a favourite picnic spot, the recent history of this Raven site is better than many; my records for eight seasons show that in only one was failure complete; in each of the other seven between two and four young are known to have hatched and believed to have fledged, five times from the low nest and twice from the high one, which was only used twice. But the fact that five of the successes were apparently from repeat layings is notable on two accounts; it suggests that many clutches are still taken and it shows how much Ravens owe to their readiness to try again which, of course, plundered Eagles will not do.

Driving on as dusk fell I had a breathtaking vision of three Golden Pheasant cocks scurrying through the dark underbrush at the forest edge. These fabulous but little-known birds were first released here over 30 years ago and have established a thriving population in and around closed-canopy forest.

13th February. The North Wind has brought new snow to the tops. Starting for a day's landscape painting in a distant glen I passed a flock of Greylags dropping to graze by the river, near the village, their wings glinting silver-white in the sharp morning light. The last five years have seen a big change in the goose population at this upper end of the valley; formerly up to 500 Greenland Whitefronts wintered here, now there are less than 100, while there are often 300–400 Greylags where there used to be none. These fields are increasingly visited by Whooper Swans—35 were grass feeding this morning—and I fear the farmer's patience will be tried when they trample his winter wheat.

I drove over the moorland road, passing a pair of Stonechats at a traditional spot where they were first recorded long before I was born. At my destination I left the car by the bridge which is famed for a fictional artist's murder. Walking like a mobile scarecrow, festooned with painting gear and rucksack, to my pitch by a swirling river, I was nearly thrown by a fox snare. Ironically many forest rangers (no longer called trappers) admire the handsome fox and know that it kills many field voles which eat the roots of young trees; but forests cannot exist in isolation from neighbouring farms and game preserves, so foxes are killed in the forest, too. The lower part of this beautiful glen is still spacious but its character is changing fast as the little conifers spread like sentinels over every brow and fold of heather. At this very early stage of afforestation Wrens and Stonechats are the only small birds which are truly resident here. Both suffer in prolonged cold spells, particularly the Stonechats, which almost disappeared from inland haunts in the 1962–63 winter. In average winters they are found in sheltered spots throughout the hills but I think they are commonest where young pines grow among deep heather and bog myrtle.

Absorbed by the complexities of tone and colour and hearing only the jumbled river-talk I was startled by the sound of Greylags overhead. A pair flew upstream, reminding me that even at this early date the feral birds have occupied their breeding places near hill lochs. The Icelandic migrants which I saw this morning will remain on lowland fields until mid-April. As I turned for home I saw the first Grey Wagtail of the year, looking translucently pale among the wet mossy stones under the bridge.

17th February. At the Flow tonight an otter came bounding along the snowy river bank to within three yards of where we were standing under the trees, before seeing us. Then it made a right-angled turn towards the river, galloped out over the broad rim of ice—which looked almost too thin too bear its weight—then dived back under the ice. Almost at once it appeared again out in the water, looking at us and making its sneezing anger call. It swam around for perhaps half a minute before moving away upstream. There is often an otter on this stretch of river and I have seen two together—once I saw a harrier diving and chattering where one had disappeared into the marsh and I suppose it is possible that one might try to stalk a roosting harrier. Eight harriers came in to roost, two males and six ringtails, rather more than I expected in this cold weather, but a thaw has begun. If there had been frost tonight the numbers would certainly have been lower. It may well be significant that in hard frost the roost is more accessible to foxes which are the chief danger to all ground roosting birds. But we do not know where the harriers spend the night when they do not come here; they are never known to roost in trees. Even if their ancestral home was steppe country, tree roosting would seem a useful adaptation in forest country.

26th February. Cold weather continues, at the High Glen the river was a dark trickle between massive ice blocks. This is the time to see Water Rails out in the open, but I was surprised to find one so far up the glen. Could there be Spotted Crakes up here as well? Down in the valley M.F.M.M. and I put one out of a ditch one February day when snow was lying—it disappeared under a bank like a rat and we could never find it again, though I watched that ditch at dawn and dusk. A mixture of Blackcock and Greyhens were on the willow buds and I noticed many trees had bark stripped from their trunks, probably by brown hares which abound up here. There were fresh mole-hills all over a field by the river; evidently the rock-hard ground does not stop their activity. An emaciated corpse of a Barn Owl was brought in yesterday; it is well-known that they are more susceptible than Tawnies to cold weather and I read that this may be due to a lower percentage of fat reserves in Barn Owls.

MARCH

1st March. The full-combed cock Red Grouse are noisy and conspicuous just now, prowling aggressively on their favourite knowes and sallying over the heather in towering display flights. Today as I walked up to the saddle below Ben Gairy the air was filled with their guttural challenges and oddly hawk-like whickering cries. For a few years the density of grouse may be greater on this ground, with its short-term flush of heather among the young conifers, than on the Moor. Competition for territories, begun in the autumn, has reached a new peak, forcing some birds on to the least nutritious ground, with little or no heather, where they are unlikely to survive to breed; research shows that such birds are more liable to be killed by predators, too.

Coming down from Ben Gairy I was alerted to the approach of one of the pair of Eagles, which nest there, by six grouse flying fast and high overhead in a wild scatter. Moments later the Eagle appeared, immense and black above the hill. A cock Merlin, tiny as a fly at this range, darted up from just below the crest and in a flash had climbed high enough to launch a series of stoops at the Eagle, always veering away before risk of contact. The big bird merely tilted a wing and dipped below the horizon again.

I could see a Raven on the nest as I passed below Little Gairy, so the clutch may be complete. A pair of Peregrines have been in residence here at least since mid-February but they may retire to one of their less public eyries before laying time.

3rd March. Clouds lapping the slopes of the Seven Mile Ridge brought scurries of snow, with brilliant sunlight between; a day to be tempted by the valley with its massed wildfowl, augmented now by returning flocks of Curlews, Lapwings, Oystercatchers and Black-headed Gulls. But there may be no better day in all the year for Eagle watching. So to the Dark Gairies again. Before the forest spread on to Ben Gairy the ground was under constant surveillance by a shepherd widely known for his fanatical love of the Eagles. They were *his* birds, ever since they had first arrived, some 25 years before, and driven the Peregrines across to the Little Gairy. Our observation point today was almost three-quarters of a mile from the big hill which looked deceptively near with its maze of rock faces and ledges mapped by sunlight and deep shadow. One of four or five likely nest sites must now be fully built up and eggs will be laid during the next fortnight. We saw nothing of the birds in the mid-morning hours and the only visible nest looked neglected, but it is hard to tell from this range. R., for whom Eagles and Peregrines are a year round passion, found fresh branches at this site last September, so we expect it may be used, but the birds have also made their customary refurbishment of the oldest and most vulnerable of all the eyries. This pair of Eagles have a bad history of disturbance, from egg collectors, photographers, sight-seers and forestry operations, so it is not surprising that they regularly change their choice of nest-site.

Soon after midday we spotted the birds soaring in wide circles over a distant ridge. As they drifted slowly nearer, a smaller, long-tailed predator rose above a belt of forest and began to soar with them. A Sparrow Hawk?—but no, though dwarfed by the Eagles, it was too large even for a big hen Sparrow Hawk and the rounded wings seemed both long and broad, a Goshawk surely. It soon turned away and flew out of sight with a few powerful wing flaps between short glides. The pair of Eagles were continuing their high, slow approach when the hen bird, recognisable by her larger size, suddenly plunged into a slanting dive at breathtaking speed, braking only at the last second as she threw her wings forward to alight on the massive nest pile which had earlier looked so unpromising. She stood on the edge of the nest long enough for us to focus telescopes and admire her silver-gold nape, then settled into the deep cup where only the movement of her head betrayed her presence. While she worked at the nest the cock had settled on a rock far above her and remained there for two hours. Meanwhile the hen, lifting effortlessly, began to make short flights, landing and walking clumsily up a grassy slope, plucking bunches of wood rush which she carried in her feet to add to the nest-lining. At times the eye could follow the movement of her dark shadow across the sunlit rock faces more easily than the bird herself.

There was no need to approach closer; the eagles had revealed their choice of nest and we had had our fill of visual delight before the snow began again and the entire hill vanished into a white cloud.

10th March. Evening at the little loch above the village; slushy snow on the ice and the threat of more snow from a sky like a wall to the east. Even so, silent flocks of Skylarks were steadily passing towards the hills. No sound from a huddled pack of Lapwings sheltering among rushes at the ice edge. Four meandering lanes through the slush led to two pairs of Mallard, their orange and vermilion legs glowing against the grey background. I waited till well after sunset in searching cold to count the Carrion Crows going to roost in the old birch wood. 101 passed in half an hour, mostly in obvious couples, but some in straggling parties of six or eight. Flight lines suggest that most had spent the day on lower ground, much of which is keepered! One bird showed traces of Hoodie grey on mantle and underparts—these 'hybrids' are puzzling as they appear to make up a constant, though very small, element in the local population while pure Hoodies are virtually unknown here, and further north are reported to be retreating as the Carrion advances.

14th March. Warmth in the midday sun at last and a beguiling, soft light on the Seven Mile Ridge, now only flecked with snow. There has been no growth of green grass during the cold spell; unploughed fields have the pale ochre look of an Italian landscape in summer, emphasising the metallic brilliance of the blue river. The pleasure of a day at my easel in the long field by the river-bank owes much to the sight and sound of paired Lapwings, Oystercatchers and newly-arrived Redshanks. The last have made a 'leap-frog' migration, over the heads of Icelandic winter visitors which still throng the estuaries. Greylags, undeterred by my presence, splashed down on the river to bathe and even began to graze out into the field where I was working. Six Bewick's Swans had joined the Whoopers two fields away; March is the month when Bewick's appear here most often. All day Rooks were carrying sticks to the little rookery at the farm on the hillside. This rookery is something of an outpost, there being none in the hills to north and west for many miles, but Rooks from here will range widely over the moors during the time of summer food shortage. Cleaning my palette at the riverside I stopped to admire a party of newly-returned Pied Wagtails running among the first yellow heads of coltsfoot on a dry shingly spit.

At dusk a Woodcock had begun its regular roding flight over the main street of the village and the night was full of urgent spring sounds—Lapwings crying and wing-throbbing, Snipe drumming, the insane piping of Oyster-catchers, bubbling Curlews and, now and again, the peculiar 'kuk uk uk' call of a patrolling Barn Owl.

21st March. In mild grey weather the High Glen teemed with life. There were several Short-eared Owls, tumbling like Lapwings and barking in combat. One perched on a stone dyke, head swivelling, lemon eyes gazing fiercely upward at a passing flock of Meadow Pipits. Plumage colour varies greatly, some very pallid and drab, others richly tawny—as if dimorphic. Private forests are fast replacing sheep farms here and no doubt it will be good for the owls for a few years. I trust that it will be worth nobody's while to clear away the willow scrub, but it may need a positive move to safeguard it. The young twigs now are deep yellow and red, sprinkled with palest lemon and white flowers. A scarlet-combed Blackcock, glossy blue from nape to rump, was displaying beneath a willow bush to five Greyhens, threading his way among them, his lyre-shaped tail feathers coiled back like springs.

At the head of the glen I went back into winter as I climbed to the cleuch where Raven and Peregrine nest. The steep dun-coloured slopes of these quiet hills are ribbed and puckered like elephant hide. The Ravens' first attempt at building on the Peregrine's grassy ledge had been abandoned in favour of their usual rock, lower down; neither species has a regular alternative site here. While I was up at the Peregrine's plucking stance, the incubating Raven came off the nest, revealing five rather beautiful blue eggs deeply cradled in wool. The Peregrine had killed a Woodcock, as well as the usual light-coloured pigeons, and some feathers looked very like Ring Ousel, perhaps the only one to have arrived so far, since I saw none in this favourite nesting haunt today. Only one Peregrine showed itself.

On the road home I saw my first Wheatear, perched buoyantly, head to wind, on top of a stone dyke. The grey back looked powdery, almost violet, against a background of burnt, blue-black heather sprigs. One year an interested postman saw the first Wheatear on 11th March but the last week is an average time for arrival.

27th March. To the Flow for a last watch before the harrier roost is deserted for the summer. Scores of frogs on the forest path, Blackbird and Song Thrush singing from the conifer tops, Tawny Owls calling and Woodcock grunting and lisping overhead. Even the Flow itself has sprung to life with song-flighting Curlews which nest in the drier part. Distant crooning of Blackcock at a lek mingled with children's voices from the farm. We were relieved to find that the crofter had not burnt the grass on the Flow—once he did and I feared the harriers would never return but the grass was long enough again by autumn. Only two ringtails and one grey male came in, though conditions were stormy. We had hoped for the late March peak of ringtails which often occurs, particularly if weather turns stormy after a mild spell. Pairs have already been displaying over their breeding territories, and the weather has probably not been severe enough to drive them back to the roost at night. It is still a mystery whether the birds at the winter roost are mainly the local breeding population or winter visitors to the region.

Perhaps the most interesting by-product of these watches has been the discovery of a spring build-up of Goldeneye roosting communally on the forest loch. They feed up-river by day and we see many small flights coming down to the loch near sunset. Tonight a raft of 50 were clearly visible in the moonlight. They always gather in the centre of the loch, in the safest place.

APRIL

2nd April. After a night of hard frost a Chiffchaff sang in early morning sunshine by the river and several Sand Martins had arrived. Goosander pairs are working up the glens towards nesting sites now. They spend much time out of the water, resting and preening on shingle banks, revealing the deep orange hue on their underbellies and their brilliant vermilion legs. Watching one pair closely I noticed that the sheen on the drake's head, even in bright sunlight, is dark green, bronze or purple, entirely lacking the sharp emerald on the head of a nearby Mallard drake.

In the water the Goosanders hunted for fish by submerging their heads to the eyes; once the duck, proceeding like this at speed, was visibly disconcerted by bumping her crown against a rim of ice under the bank. Sun turned their wakes to a sparkle of gold; suddenly the duck was still, her body flattened and tail cocked in the soliciting posture, but she seemed unregarded by the drake at present. In a patch of old bracken screened from the road by a stone dyke I stumbled on a Mallard duck incubating seven eggs and stopped to make drawings of a sitting Woodcock, on a nest of four eggs. She sat perfectly still, one enormous dark eye watching me, her barred, grey-tipped tail inclined upwards, her entire plumage a marvel of camouflage in the warm brown of old bracken and paler litter of last year's oak leaves.

7th April. Up on the moor Jimmy, the herd, was burning heather and the boss, who owns several sheep farms, was with him. There is much talk that he may soon sell the Moor and then the forest will take over. He is friendly but predictably non-committal on future prospects. Over the next few weeks Jimmy will be totally absorbed with the lambing.

I am always fascinated to watch the hazy smokescreen of moor fires, as they seem to engulf whole hillsides. Sometimes, out of control, they raise bitter controversy between game preservers and shepherds. The systematic rotation of burning practised on the big grouse moors is not found here. In the previous herd's time a great tract of old heather was burnt one spring on the Lion's Head, as I christened the rugged hill which is the most impressive feature of the Moor. For years this had provided wonderful nesting cover and the year before it had been the site of my first Hen Harrier's nest in the district. I shall always recall the early April day when I found it charred and bare of all cover and my astonishment at finding that, even so, a pair of harriers had returned to it. What a dazzling, pearly vision the cock made as he floated slowly across the blackened slope and landed beside the brown hen not 50 yards from the old nest site! So strong was their attachment to the home ground that they moved only a minimum distance to nest in a small patch of long heather some 500 yards across the burn.

The Moor is still important to harriers as a hunting ground but in more recent years nests have most often been sited in the dense ground cover of the neighbouring young forests.

A pair of Golden Plover were on recently burnt heather. Through the glasses the spangled markings of their upper plumage are amazingly intricate, almost impossible to resolve into individual feather patterns. The cock's face was black, the hen's mottled yellowish-brown. The number of breeding pairs on the Moor fluctuates between 10 and 20, with two or three always quite near the road, on flat or gently sloping ground, short heather interspersed with damp sphagnum hollows and drier mounds, but the density of pairs is highest beyond the little ridge, around the margins of the boggy Fleshmarket. Up above the herd's house I could see the Lapwing pairs, dancing like puppets in the sky above the grass fields where some clutches are already complete. A few pairs, tending to form new colonial groups, are increasingly found further up on to the open moorland. During an afternoon shower of rain I had a wonderful close look at a cock Lapwing, with the rain-drops gleaming like tiny pearls on his sleek, glossy back.

8th April. Two Swallows were seen in the valley this morning but in the afternoon it was all cold rain and low cloud in the hills. When the sky lightened briefly we saw the cock Eagle leave the cliffs of Ben Gairy and salute his sitting mate with a plunging dive; suddenly a straggle of 30 Pink-footed Geese passed not much below him, heading north-east, and I momentarily caught my breath, thinking of tales of an Eagle striking down a goose from a flock migrating over the Cairngorms. But the little skein passed safely on, soon lost in cloud again. Most mornings now, in the village, I hear or see Greylags or Whooper Swans as they pass north up the valley, with Iceland their ultimate goal. At the Little Gairy we had a brief but satisfying view of the Peregrine tiercel and heard the falcon's strangled food call but could not spot her in the mist. A cock Ring Ousel with shining white chest bounced on to a road-side dyke, and later we heard his desultory piping from the invisibile crags above.

11th **April**. L. has seen a cock Hen Harrier displaying over the forest beneath the Seven Mile Ridge, and two ringtails in the vicinity. We have not found a nest there in past years so decided on a morning's watch. Before starting out I saw a pair of Siskins from the kitchen window, feeding on peanuts. This habit, new to the district, has been observed at three different spots in the last fortnight. This could be the time when Siskins find 'wild' food hardest to come by.

We watched the forest between ten and one o'clock. A pair of harriers showed up almost at once, close circling. After a short disappearance the cock returned alone and made a diagonal display run, crossing the whole face of the forested hill. He rose and fell in the sky in a frenzy of flailing wings, rolling on his back at the top of each steep climb. Then he drifted away, to return and repeat the display three times more in the morning, sometimes varying it with shallow switchback dashes. This performance has been aptly christened sky-dancing. Later there will surely be a nest somewhere beneath the line of these flights.

Several Short-eared Owls were hunting and some certainly have mates incubating eggs. Nests are difficult to pinpoint in the vast undulating panorama of trees. One owl appeared to drop a vole to a sitting mate not far away, approaching so low between the trees that it was impossible to mark the spot accurately. A brief search proved fruitless. Perched on top of a Sitka, this owl looked slim and tense, with ear tufts conspicuously erect. Later it rose with deep slow wing strokes and we heard the muffled 'hoo hoo hoo' of its song from high in the blue sky behind us and watched the bird collapse, as if shot, and fall with closed wings cracking together beneath its body.

In this forest, which is not owned by the Commission, deer control is very severe and even Blackgame dare not show their heads, but owls and, I trust, harriers are valued. This is the season when Red Deer do most damage to the bark of young trees. Ian, a Forestry Commission ranger, thinks that this is not a straightforward case of food shortage, though it may be due to a mineral deficiency.

14th April. Annually at about this time Derek and Chris are welcome visitors, inspecting Peregrine eyries. There was no sign of life at the Little Gairy after all and D. thought the wool hanging from the now deserted Raven's nest indicated that it had been robbed. We looked up a narrow glen where young Peregrines had flown once before from a site above a steep heather slope which looks more suitable for a Kestrel, though in fact it was an old Buzzard's foundation which the Peregrines had appropriated. They had indeed cleaned out a scrape ready to lay this year but they did not appear. D. carries an encyclopedic history of many pairs in his head and I have learned much from his interpretation of behaviour. Some pairs have full clutches by now but in these days of thin egg-shells they may disappear quickly, broken and eaten by the falcon; later a repeat laying might be successful at a different site. In afternoon warmth an adder, two lizards and a peacock butterfly were out.

21st April. Last week's watch on the harrier pair raised speculation on the chance of an early nest. A pair in north-east Scotland broke the record recently by fledging young on 12th June, which meant that eggs were laid in mid-April, at least a fortnight earlier than I have known them here. We were up on a heather slope above the forest early this morning, rather too exposed for comfort and perhaps for the harriers themselves but they appeared perfectly at ease. From this high point we could observe their frequent landings and note the abundance of likely looking nest-sites. There was no more sky-dancing but the pair were fascinating to watch as they weaved slow patterns in low flight among the trees, tails fully expanded in courtship, but we saw no carrying of nesting material.

The influx of breeding birds to the plantations is noticeable now. Reed-buntings which a week or so ago were in small flocks at the forest-edge, are singing in territories, Meadow Pipits song flighting all around us and Willow Warbler cadences drifted up from below. Coming down the glen road, a cock Redstart flashed and a Common Sandpiper was on the river shingle. A Dipper's nest on the buttress of a bridge caught my eye; it is ten years since this particular site was built on.

25th April. Last week egg-collectors were chased away from a Peregrine eyrie and now I have been told the car number of another collector 'who visits the area at week-ends'. Judging his likely route of entry I watched the road from six o'clock this Sunday morning. Just before nine I was thinking of returning home when he flashed past. I turned to follow him but could not gain another sight of the car. A visit to the police station ensured that most likely spots would be watched. We searched till dusk but no one saw the car again. Now I think I know where he went and he has probably bagged at least one Peregrine clutch.

The first departing flocks of Greenland White-fronts have flown north over the village on this fine warm evening; the rest are sure to go in the next few days. I am unashamedly sentimental as I watch them go. For six months their laughing cries have belonged to the valley and the hill lochs.

MAY

4th May. An early start brought an unexpected reward as we drove beside the big loch en route for the Moor. A dark shape on a partly submerged tree stump was not the usual Cormorant, but an Osprey. It took off but we followed it back to the river entry, saw it plunge and rise clasping a perch which it carried to eat on a convenient post. It had gone when we returned in the afternoon.

When we arrived at the Moor it was already as hot as a summer's day. We had a talk with Jimmy the herd. He has lost some lambs to foxes and Johnny the ranger is coming tonight to help him tackle the den which he found at the back of the Lion's Head. Jimmy has been seeing two different cock harriers hunting over the Moor. He says both head back into the forest with prey. The flight lines he describes suggest that pairs are in occupation of two of last year's nesting territories. It is interesting that the cocks come over the Moor so much, even sharing the same hunting grounds. No hen has risen to mob Jimmy on his rounds so there is unlikely to be a nest on the Moor. We took a stroll up on to the little ridge and it was pleasant to overlook the familiar scene once again. Colours are still predominantly the browns and ochres of winter, the curled heads of fresh green bracken only showing at close quarters. Reddish-purple stems of bog myrtle, still leafless, are clustered with orange catkins. A big brown moth, gone in a flash, must have been a male Emperor, already flying in search of the beautiful grey female. Just below us a cock Wheatear sang wheezily in butterfly flight, spreading his black and white tail, and soon a hen flipped out from under a tumble of rock where the nest must be hidden—had we walked on without stopping she would not have come off the eggs.

I found myself wondering again about that strange discovery, one May day, just down the slope from here, which strongly suggested that the female of a harrier pair had been killed by an Eagle. We found her dead body lying face downwards on the heather, with wings outspread. A ring on her leg identified her as one of a fine big brood of five nestlings which I had ringed in a patch of purple moor grass down by the Fleshmarket, only a moment's flight from here; she was almost three years old when we found her and had obviously returned to breed in her natal area. The only mark on her body was a wound just below the sternum and the puzzle was how had this been inflicted? Beside her lay a very big pellet, containing the remains of Red Grouse and Blue Hare, too big to have been cast by the harrier herself. Later I sent it to the late Ernest Blezard who agreed that it appeared to be an Eagle's pellet. At first we thought the harrier might have been shot but a post-mortem by Ian Prestt revealed no trace of an exit hole and he agreed with my hunch that the wound could have been made by an Eagle's talon. I could imagine what had happened as I had seen the rehearsal the week before, the pair of harriers diving at a pair of Eagles. Next time, perhaps, the bold female harrier (her mother had been a ferocious character) had dived too close to an Eagle. What a drama to have witnessed! I cannot prove it but I believe it happened. The mystery deepened when we found a despoiled grouse nest a few yards away; this we reckoned was the work of a fox from the evidence—the wing of a hen grouse bitten off, a trail of feathers and eggs chopped through and partly eaten. Perhaps the fox was involved in a triangle of predators—but what was responsible for the plucked wing and breast bone of a Snipe lying beside the dead harrier we could only guess. Most likely it had been harrier prey.

My best memories of the Moor go back to the early years, when I first discovered its charm, finding exquisite Golden Plover chicks when it had seemed impossible to outwit the parents' tireless vigilance, or learning how Curlews, by furious attacks, can deflect sheep from trampling chicks. My awareness of the birds and the scene, in varying moods, sharpened over numerous visits. The breeding harriers were the focus of watches at all hours of the day. In grey early light I watched a cock leave his perch on an old fence for his first hunting flight and saw him return, often between seven and ten o'clock in the morning, with food for the hen. It was always a thrill to watch her rise for the food pass, the light catching her barred grey underwing as she swept over the dark heather to meet the cock as he glided down. Those early mornings were full of Curlew song and the calling and chuckling of mating Cuckoos. The midday hours were much quieter and I no longer heard the song of Whitethroat and Reed bunting from the willow bushes by the burn. I learned to listen for the swearing crow which often announced the approach of a cock harrier or sometimes of Eagle, Buzzard, Peregrines or Short-eared Owl; a pair of the latter often nested near the harriers. For long hours the cock harrier was far away, but a food pass could always be expected again between five and six o'clock in the afternoon. During incubation, two feeds a day were usual for the hen. Between these she never moved from the nest deep in the heather.

How to evaluate the charm of this country? Ecologists say it is a degenerate habitat, overgrazed and overburnt. If it were not for sheep it could be natural forest again or, failing that, planting conifers would increase the population of birds and other wildlife. But I am sceptical of quantitative comparisons. Ask the hill walker or the shepherd and both will say there is more joy in life when they can hear the songs of Lark and Curlew, feel the spaciousness of the open hill and climb a little nearer to the sky. If, as has been suggested, the Moor is worth preserving as an open space, in surrounding conifer forest, the problems of reconciling human recreation and wildlife conservation will be testing.

10th May. Up at the head of the High Glen the Peregrine's nest was empty. The falcon has probably eaten the eggs which were laid in April. I settled myself lower down among the rocks to watch and draw Ring Ousels for an hour or so. A male was piping from the sheep stell below. Soon a hen flew past and disappeared into a thick mat of blaeberry on the side of a steep little cleft. She came out when I went near and I was just able to see into the cup of the nest which contained four eggs. The herd likes the bold and lively Ring Ousels and remarks regretfully that there are fewer this spring. Often there are three or four pairs within half a mile of his house, including one nesting in the stone dyke at his garden—there are no Blackbirds up here. The Ravens' nest had three well-feathered young, huddled against the black overhang of rock. Their wings are already glossed with blue-green and they have strikingly pale cloudy-blue eyes. The herd has had no trouble with his flock from these Ravens and the young should fly safely; one year a fox took a younger brood at this nest. In the same gully as the Ravens I noted the usual pair of Pied Wagtails nesting, and a Wren singing; after the hard winter of 1963 both were absent. I found a Kestrel with five eggs on a ledge which I could almost reach from the grassy slope below.

The willow thickets in the high valley are now alive with small birds. The pair of Willow Tits probably have eggs in their nest in a rotten stump and were very quiet, but many Redpolls, Willow Warblers, Sedge Warblers and Tree Pipits were singing. Several richly marked cock Reed Buntings and one or two brilliant Yellow Hammers sang from the tops of the bushes. In the past this glen was always outstanding for both Whinchats and Wheatears. Now the Wheatears are declining; no species is driven away more quickly by the spread of afforestation which very soon makes it difficult for them to find sufficient short turf or bare ground for their feeding habits. Whinchats, on the other hand, take much of their food among tall grasses and other plants. Apart from the willow scrub and some shelter belts of conifers, there is very little old woodland here; a neighbouring glen with birches and alders along the burn has Redstart, Pied Flycatcher and Wood Warbler which I do not expect to see in the High Glen. On the way home I looked in on a Goosander sitting on ten eggs in a hollow rowan tree, which has been occupied many times before. Once known, a site is worth a look every year until the tree falls or the nest hollow becomes too exposed. Once, in this locality, young were successfully reared in a castle ruin. I wish I knew whether the Barn Owls which normally breed there, did so that year.

An evening visit to the Peregrine eyrie where collectors were foiled last month confirmed that the original site, where there was no proof of egg-laying, is now empty and deserted. The falcon flew into a new site, where she must have eggs, further along the cliff.

12th May. Afternoon watch on the harriers in the forest under the Seven Mile Ridge. We found a position with the sun behind us, across a valley from where we had seen the cock displaying. We had hardly settled when he appeared with prey in lowered foot and the hen rose for the pass, from a block of young pines near a ride. The nest must be at a height of 1,000 feet, unusually high for the district. Before she returned she gathered a large bunch of grass in her bill, evidently still adding material to the nest, though she is obviously on eggs. Later, while the cock was flying near the nest, a second, browner, male joined him; for a moment the hen came off and all three circled together amicably. Perhaps the visitor was the cock of the pair said to be nesting beyond the Ridge. A. and I had a day on the tops last week. They were birdless except for a very few Wheatears, Larks and Meadow Pipits, a pair of Golden Plover on the highest summit and a Ring Ousel singing in the corrie just below. While we were up visibility suddenly became almost nil, with low cloud racing across the plateau and we had an interesting half hour trying to decide which way to come down.

16th May. Out along the moorland road to the Dark Gairies. Stopped to make sketches of Whinchats, irresistibly fresh and vibrant subjects on the roadside fence. A first brood of newly-fledged Stonechats were being fed close to one of the cock Whinchats. Suddenly the cock Stonechat flew at the cock Whinchat and chased him away, but he soon returned, singing. His song seemed exceptionally rich and varied. I caught the rippling notes of a Whimbrel overhead—the northward passage is usually at its peak in mid-May. Reports that the Ben Gairy Eagles have failed proved sadly true. No eaglets have hatched and a deserted egg, taken under licence, has been sent to be analysed for toxic chemicals, but human disturbance is the likeliest cause of failure. The pair were circling and diving around one of their alternative eyries but they will not nest again this year.

20th May. A keen north-east wind with stinging hail showers, and sleet at times, has put the calendar back to March. I made a long journey to a distant rocky hill with a view across a moorland valley to a small clump of trees on a bracken-covered slope. Wild hyacinths blooming among the young bracken on the open hill show where more woodland used to be and I found the first bright pink lousewort flowers on a damp slope. Sheltering in the lea of a long buttress of rock, I watched the sky for a cock Eagle returning to a nest in an old pine. Last year, just as the eaglet fledged, the huge nest collapsed to the ground, pounded and saturated in a summer storm. Is there a new nest in the same or a different tree? Between showers I made sketch-book notes of a wonderful, fleeting effect as a distant hill emerged from cloud, freshly whitened with snow.

For a long time the trees seemed lifeless. At last in mid-afternoon I spotted the cock Eagle soaring above the far crest. Just then a flight of homing pigeons began to cross the hill and, as they passed close to the wood the Eagle came behind them with extraordinary speed. For a moment I thought the pursuit was serious but as soon as they were clear of the trees he rose again and circled out of sight. It was an hour and a half later that I spotted him making a high, slow approach, carrying prey. A passing aircraft bothered him and he retired again. Half-an-hour later he was back in the sky, still carrying the bundle of prey. Was he bringing food just for the sitting hen or was there an eaglet to feed? This pair are normally a week or two later than the Ben Gairy birds. He swept down past the corner of the wood, with wings half-closed, and probably alighted at the nest which would be hidden from my view by nearer trees. A moment later an Eagle flew low among the trunks and pitched in the grass at the wood-edge. I knew this was the hen because of her plumage colour; remarkably, she has had a white tail base, generally a mark of immaturity, through two successful breeding seasons and even now it is still fairly notice-able. I could see her tearing at the prey. Twenty minutes later she flapped heavily back to the nest tree and the cock came out and glided away over the bracken. Almost certainly the new nest is in the old tree, but I decided to wait a week or two before taking a closer look. I would judge that the chick or chicks, if hatched, are very small.

27th May. I have heard the Forest Loch compared to a Finnish Lake. The comparison seemed nearer reality with a late pair of Goldeneye on the water.

One day, perhaps, they will breed here. Alas, only a few of the old hardwood trees, mainly birches, now remain in the twenty year old forest further along. Still, the mixture of trees and the uneven growth of the conifers, due to heavy deer browsing in early years, make this a rich area for woodland birds, including Green Woodpeckers. I passed the isolated croft with its throng of children and delightful enclave of green pastures; Lark song and the bleating of Snipe overhead, Lapwings with chicks and an Oystercatcher sitting in the grass. A pair of Redshanks 'chipped' from fence posts at the farm cat. Nine Blackcock were leaving the larch tops, where they had been feeding, to gather for the evening lek. On the river seven small Goosander ducklings swam in a bunch behind a duck and began to scramble on to her back. A duck Teal had spotted me and sat perfectly still under the bank, shielding her brood. I decided there was time to walk up a long slope with young trees to the deep heather and rocks above the forest in the hope of a Merlin's nest. One very accessible Merlin site is well-known and it is a wonder that young are reared when visitors have beaten a path to the nest. It would be exciting to discover a new eyrie and add a piece to the jigsaw of local knowledge. A long scaur of rock, with many broad shelves of dense heather and blaeberry looked a likely spot, the more so when a cock Merlin rose at a Carrion Crow which came off a nest with eggs in an old rowan. When the hen Merlin rose at my feet in a flurry of wings and barred tail I felt elated and guilty at the same time and stopped only for one satisfying gaze at the five pale red eggs on their peaty bed.

Coming down through a jumble of fallen boulders, to the larches again, I saw movement below and found myself looking down on a vixen, lying on a flat rock, with her back to me, surrounded by four romping cubs, one tugging at her tail. I sat still for an hour, watching and sketching them. Fortunately the wind was right and they were never aware of me.

Long after sunset I stopped at a broad ride in the older forest to hear the first churring Nightjar of the year. Woodcock were roding and young Tawny Owls calling.

JUNE

1st June. On a fine sunny morning with light east wind I set out for the Seven-mile Ridge. By evening thunder clouds often gather. On the way through the forest I saw a very beautiful big fox, looking like a Liljefors painting as it stopped in its tracks, orange-yellow on the body with grey tail broadly white-tipped; colours glowing against the dark edge of forest. Up the steep slope above the last trees a Meadow Pipit rose off a nest with four eggs, under a tussock of moor grass. G., who finds scores of nests in the forest, has seen no clutches of five this year, perhaps correlated with generally poor weather. I passed a big black rock face, where a Ring Ousel sang, a good looking place for an Eagle; but there has not even been a Buzzard or Raven on it recently. A keeper friend is always complaining that the local Buzzards are increasing; L. saw eight together in April, at the head of the next glen, but where are they all now? Crag-nesting pairs are rather sparse, but there are more nesting in the old plantations around the fringes of the hills. There seems little profit in protecting them from collectors in the spring, when the young are so often shot as they leave the nest! A short climb and I was on the dry, springy turf of the ridge. On a day such as this walking here is exhilarating and peaceful. The colours of the ground, carpeted with blaeberry, dwarf willow and cladonia lichen, are quiet and very various; here and there are conspicuous little tufts of fir and alpine club-moss, like miniature conifers. On the summits the ground is barer, with many light-coloured stones and on the declivities the vegetation quickly becomes grassy and quite lush. Compared with many High and tops the extent of alpine ground is small, but in times past this was Ptarmigan country, though none have been known here for well over 100 years.

My old friend, the Rev. J. M. McWilliam commented, years ago, when we first walked the tops of this region: 'You will not find Dotterel because there are none'. Once, not so very far away, I did have the wonderful experience of finding a nest but I regard this as a memory that must suffice for a lifetime. This pair of Dotterel chose a gentle grassy slope, not much above 2,000 feet, for their nest. Somehow I never expected that nest to survive—it would be trampled by sheep, a fox would find it or the incubating male would be swept away by a Peregrine when he came off the eggs to feed and run conspicuously over the grass. I was as suspicious of predators that summer as any game preserver. How illogical can one be! No Dotterel have returned to nest near the spot, though two chicks were successfully hatched and almost immediately taken 1,000 yards away by the male to a stony summit where there was abundant insect food, a remarkable journey, apparently necessary for their survival. If there was still a female Dotterel on the hill after the male began to incubate, I never saw her.

But to return to the Ridge—there were, of course, no Dotterel, but the solitary pair of Golden Plover on the highest summit complained as if they had chicks in hiding. Over all the top I saw only one Skylark. I wish that I had counted the larks on every summer ridge walk; as it is, I can only surmise, from my notes, that they have markedly decreased over the last decade or more. I used to note them as rather commoner up here than Meadow Pipits, though neither were plentiful. Is a decrease of Larks up here a reflection of their much more drastic decline on lower hill ground due to afforestation, or should one expect an *increase* on the tops with the loss of that habitat?

A Red Deer hind, with two younger animals, probably her calves of the previous two seasons, was grazing the grassy slope on a spur below the highest summit. She looked ready to calve again and is probably being watched by the forest rangers who are tagging the new calves this week. I saw a single blue hare, not quite changed to summer brown; outlined against the sky the short ears and blunt rounded nose very distinctive.

2nd June. A message last night from a forest ranger that he had seen a bird, like a small bittern, at a little loch near the Dark Gairies. A search for it this evening proved fruitless, but the description leaves almost no doubt that it was a Little Bittern. The spot looked unlikely enough, with only one thin patch of reeds where the bird was seen. But it wouldn't be the first rarity thereabout—it was on a forested hill nearby that, many years ago, the extraordinary discovery was made of a pair of Eagle Owls displaying. How often can precious time be lost in following up reports of strange birds, though they may lead to new contacts among interested people. Last week a lad working in the forest led me miles after a 'Greenshanks' nest' which proved to be a Common Sandpiper's. I was sorry to disappoint his obvious keenness.

At the Little Gairy, the Ravens had two big young from their repeat clutch, in the high nest.

5th June. Yesterday L. found the Tufted Duck's nest with nine eggs by the little loch above the village. We have often seen a pair there in May but have not seen a nest before, though a good number breed in the marshes down the valley. On some of the larger lochs broods are increasingly disturbed by power boating and water-skiing. It is interesting that last summer and now this, nests have been found at these new sites beside very small, relatively undisturbed lochs. Back in 1893 John Paterson searched most of the lochs in the region for breeding Tufted Duck and reported only a single pair on one loch. Shortly after this date they had colonised many of the lochs where they have since bred, mostly in the lowlands. They are still absent from most of the hill lochs which look very similar to lochs where they breed in the Hebrides. In 1897 three-quarters of the young were taken by pike and there are still a lot of pike in some of the breeding lochs. A Corncrake has been audible from the village the last few nights; this seems to have been a better year for them, but there are very few still compared with twenty years ago.

We decided to make the journey to the tree-nesting Eagles, this time to find out for sure if there is an eaglet there. We approached the wood about five o'clock in the afternoon. Very quietly we edged into the trees. At seventy yards we were looking up at the dark mass of the nest, rebuilt as before on the first branches of the old Scots pine, thirty-five feet from the ground. Partly shadowed by the higher canopy, the hen Eagle was standing on the side of the nest, with her back to us, a white eaglet watching her alertly. Incredibly neither was aware of us. The hen was feeding the eaglet, delicately passing it small slivers of meat with her huge bill, from what looked like the carcase of a hare. A thought flashed through my mind—this is like one of those shots in a colour film that the photographer waits hours in a hide to get. At last—it was only a few moments later—the hen eagle shifted her stance, stood for a moment in full sunlight and launched herself full-face directly towards us. She was overwhelmingly close before, seeing us at last, she banked swiftly away and sailed out across the valley. We went forward to examine the tree and look at the nest. The litter of branches where last year's nest had fallen still lay below; the new nest looked a hurried structure, as it must be—the old nest had been built up over four seasons. The eaglet was nearly a month old. We found one unexpected item of prey under the tree, the hindparts of a recently fledged Short-eared Owl. But the most alarming discovery was that someone had left the marks of climbing irons on the tree. I have an idea who may have done this and he would have a licence 'to examine the nest' but I fear this may mean that disturbance will eventually lead to the birds seeking another site. Later we watched from across the valley for two hours. For much of the time the Eagles were flying all round the sky, circling and diving in play, like a pair in courtship; their flights, as always, were excellent indicators of other species on the hill; most interestingly a cock Merlin rose to mob them from a likely slope for a nest, which we failed to find. Once a Common Gull flew cheekily close to them until they suddenly turned on it and sent it away squawking.

12th June. We have been trying to get up to date on harrier nests. At the end of May we watched for half a day in the forest under the Seven-mile Ridge without seeing a food pass. Finally that evening we decided to go to the nest, thinking it must have failed. At close quarters we had the usual difficulty in identifying the exact spot where we had marked down the female Hen Harrier on 12th May. We were almost despairing, especially as we had so little hope of an occupied nest, when she rose, yellow legs trailing, just ahead of L. The nest, containing four well-incubated eggs, was typically placed in a rather open patch of lodgepole pine. The female was moderately aggressive and came close enough to show that she was probably a year-old bird, from her dark brown iris colour. Recently the hens of several breeding pairs have been one or two year-olds, but I have only twice seen a breeding cock in first-year brown plumage. Unlike the Orkney situation where some cocks have harems of several hens, our cocks show little tendency even to bigamy. We do not seem to have any excess of hens over cocks, possibly because hens are more vulnerable where there is a lot of human persecution. The high proportion of young hens breeding, of course, suggests they have a short life expectancy. Yet, by sexing nestlings, I have found that almost twice as many hens as cocks are consistently fledged from local nests.

Today we looked at three areas which have held harrier pairs before, in the forests near the Flow and the Moor. At the first, where the trees are possibly too dense now, we drew a blank. At the second, after a long watch from two positions in difficult uneven ground, we at last found a nest with three small young, in knee-high heather on the side of a hollow. The cock, I think, is the fine old silver-grey bird which has been here for several years but the hen is new, with the amber eyes of a fairly young bird; she is much fiercer and darker than the old light-breasted hen which bred here for three years. The only prey in the nest was a fledgling Meadow Pipit. While watching we saw a magnificent Red Deer stag in velvet and glimpsed a fox which the female harrier had been diving at. She also had a tussle with a hen Sparrow Hawk which must have a nest in the taller larches up the slope. In a few years all this great spread of conifers will be Sparrow Hawk country and the harriers will probably have moved to younger plantations. At our third area we were lucky to see what was probably the last food pass of the evening which showed that the nest is about 400 yards up a slope from last year's, which was on level ground and very difficult to pinpoint among the trees. It was interesting to see the cock bring prey during heavy rain, which decided us against inspecting the nest tonight. Earlier in the evening L. heard several Grasshopper Warblers which are generally inaudible to me now. A Song Thrush almost fooled us mimicking the harrier's food call in its song. There were Redpolls everywhere, indeed there is a wealth of small birds in this young forest, but the Red Grouse which were so plentiful two years ago are very much scarcer now that the trees are suppressing the heather.

20th June. From the village I can see the scar on the opposite hill where a hardwood is being felled. In the short run these summer fellings are catastrophic for many birds though miraculously a brood of seven young Redstarts survived in a nest which was left naked, a foot from the ground. In the long-term the loss of hardwood in the region is now very serious, but it is heartening that the Forestry Commission are preserving a notable old wood of mixed oaks and birch. At four o'clock this morning I went to the Goosander's nesting tree, in the hope of seeing the nestlings leave; they began to hatch three days ago but were still there yesterday. The duck allowed herself to be lifted gently so the ducklings could be seen—how beautifully marked they are! They were gone when I arrived this morning, all nine of them, leaving one addled egg in a soiled bed of grey down.

26th June. Midsummer past, with little warmth in the sun. The Moor has become a patchwork of greens where bracken and grasses grow, heather is all shades of brown and bronzy-green, light-speckled with buds; some plants of bell heather are already crimson and there are many delicate pink heads of cross-leaved heath on the wetter parts of the moor. Damp hollows and the edges of drainage ditches are starred with pale green rosettes of violet flowered butterwort. A closer look at the ground reveals round and long-leaved sundews too. Near the old larch stell I took a nostalgic look at the Hen Harrier site of ten years ago: the grassy knowe where the cock used to sun himself is bright with the small moorland flowers—white heath bedstraw, yellow tormentil and deep blue milkwort. Here a brood of young Wheatears had lately fledged and a pair each of Whinchats and Reed buntings, sparring a little, were carrying food to young still in their nests in a patch of bog myrtle. Down by the Fleshmarket a pair of Curlews circled me, chortling explosively; one of their big leggy chicks caught my eye before it took cover. Some Curlew families have already left the hills and the Moor is becoming a quieter place again, although the melancholy piping of Golden Plover with chicks shows no sign of lessening. A Buzzard rose ahead, leaving a litter of broken-up sheep droppings where it had been hunting for dung beetles. Already the summer flocks of Rooks are foraging over the hill tops.

I walked down the marshy valley to a loch, studded with islands where Common Gulls nest, but saw no Dunlin, which were always there before the forest came. I miss their purring songs and confiding, though often baffling, behaviour. A Snipe, slant-legged on a post, made a good subject for a drawing.

JULY

1st July. High summer weather at last. Along a forest roadside the best of the broom is past but foxgloves are reaching their peak and mats of wild thyme are flowering. Between the tracks of vehicle wheels some of these roads have a centre-line of brilliant birds' foot trefoil and mountain pansy. Later Bullfinches, Redpolls and Linnets will find a feast of seeds here. Surely there is no need to spray these linear oases of colour! A suntrap between steep banks was alive with butterflies, including several Dark Green and many Small pearl-bordered Fritillaries and Small Coppers.

I went up to watch the Merlin's nest found on 27th May. The adults are not at all shy and ignored me as long as I kept my head down in the heather, scarcely more than 100 yards from the nest. Prey was brought at roughly hourly intervals during the afternoon, a meal for each of the chicks, I guess. The cock is still doing most of the hunting. His approach is easily detected by his burst of sharp calling, preceded or accompanied by the long food call from the hen. All the food passes took place on the ground, the hen taking the prey in her bill, sometimes transferring it to a foot in the short flight to the nest. She looked surprisingly pale against the heather, in the bright summer light. At the nest I found five still downy young clustered together. The down is lighter, creamy-buff on the crown, sandy-coloured on the youngest; otherwise they are bluish grey, paler below. I noticed one curious feature—two had distinct greenish tips to their bright pink tongues, while the others did not. At a later stage their tongues will be bizarrely crimson and blue and green towards the tip.

3rd July. Re-visited the harrier site under the Seven-mile Ridge. No sign of the hen as we approached so we thought she might be hunting distantly. Eddie Balfour recently saw a marked hen hunting five miles from her nestling brood. But we saw nothing of cock or hen and found a disaster, an empty rain-sodden nest and the corpses of two mainly downy young a couple of yards away. Was this a case of human destruction? The dead young lying together could have been crushed by a boot and thrown down and quite a bunch of flank feathers from the hen, near by, suggested that she had been killed or wounded, but we found no trace of her body. It was all too tidy and the young too intact for the work of a fox. True, the young might have died in wet weather, as they often do if this is prolonged, but they were too small to have wandered from the nest. We suspect that some ardent harrier-hater had been here—a sad disappointment after the spring had seemed so full of promise at this new, seemingly safe site. Coming down through the fly-infested jungle we made a happier find when a Merlin led us to a hitherto unsuspected nest of four unusually pale eggs on a typical heathery bluff above the trees; a rather late nest. Further on, a Redpoll called plaintively beside me, from the top of a young sitka. Her nest, with four light blue, red-spotted eggs was not difficult to discover at shoulder height against the trunk of another small sitka, conspicuously grey-brown among the bright green needles. Few nests could surpass this for refinement of minute construction. Scarcely two inches across the cup, the material is like a microcosm of the forest. Moorgrass, heather, twiglets of larch and shreds of bracken are tightly interwoven with a few feathers and studded with pale green whorls of sphagnum moss. The soft white lining is made from bog cotton. A close look identifies some of the feathers, tiny brown vermiculated ones from a grouse and, padded into the side of the cup, some larger ones show the unmistakable black and buff barring of a Short-eared Owl. Forestry workers must find countless Redpoll nests; what a dossier they could compile on nest material and sites, laying dates, clutch and brood sizes for a start! Crossing a drainage ditch I disturbed a slim dark bird from a compact little nest under a mat of molinia, with five off-white eggs, rather heavily marked with reddish-brown. We could not see the bird again but concluded, after some discussion, that this was a Grass-hopper Warbler's nest. This elusive bird could well be almost as numerous as the Willow Warbler in the young forests.

Driving down the hill we came alongside a weasel with two well-grown kits. They made no attempt to move off, leaving the parent in a dilemma. She rushed haead dragging one by the nape of its neck. When the car made up on her she dropped it and bounded away to a stone dyke but quickly returned to collect the kit, this time pulling it along by an ear. Presumably after depositing it beyond the dyke she would go back up the road for the other. Twenty years ago weasels were much scarcer locally than stoats; now weasels are probably the more numerous. Perhaps this change is due to an increase of voles, in the forests, coupled with the reduced numbers of rabbits since myxomatosis; but rabbits have increased again lately.

4th July. The Peregrines referred to on 10th May hatched two young from their repeat laying. John, a Nature Conservancy warden, has been organising a watch on the site as some broods have been lost to falconers or 'pigeon men'. There have been an encouraging number of volunteers, but organisation in this scattered community has its problems. I watched this morning from half-past four till nine o'clock. At first I could only see the birds at intervals, between drifts of low cloud. The two well-feathered eyasses are, fortunately, much less conspicuous now they have lost most of their white down. At five-thirty the falcon flew to the eastern summit of the crag, just as the sun broke through, and stayed there for an hour. Then she rejoined the tiercel, who is indolent for long periods, on his favourite rock slab near the old eyrie. At eight o'clock the tiercel at last went off to hunt, returning with small prey in a foot at eight thirty-five. The falcon, screaming the food call, flew to take it from him on the grass above the cliff and carried it to the nest. The young then began to feed themselves. At the opposite end of the crag a brood of Kestrels were taking their first flights. I made sketch-book drawings of the Peregrines and of a pair of Curlews close by me.

7th July. A fascinating but exhausting day in the forest near the Flow and the Moor. We ringed two broods each of Hen Harriers and Merlins. Hot, steamy weather, constant warfare with midges and flies. At the first Merlins' nest (see 1st July) the youngest chick had died, as it often does in broods of five. The others, beautiful, fierce little birds, still had sprays of down on crown and scapulars. Working rapidly I filled several sketch-book pages with drawings and colour notes; legs dull chrome, cere greenish–lemon, huge dark eyes blue-rimmed, etc. I took away a handful of pellets which later showed that food has included Meadow Pipit (adult and young), Skylark and House Sparrow (both adult), young Starling and Wheatear. As expected there was no trace of grouse chicks. The House Sparrow must have been taken down by the croft. At the second nest, two miles away, the three young were almost flying. They were crouching in the heather along the ledge from the nest itself and turned on their backs instantly when handled and struck with needle sharp claws. To my knowledge there has been a nest near here most of the last thirteen years, probably every year, and sometimes on precisely the same ledge. Even road-blasting, four hundred yards away, did not upset them one year.

In spite of a general decrease on open moorland, Merlins have locally increased on heather ground in or above the young forests. Obviously they find them well-stocked with prey and nesting sites on ungrazed slopes are good. In one such area of some forty square miles I know positively of six breeding pairs and there are possibly eight, which would agree very closely with the density formerly recorded on the Yorkshire moors. Last year our only known nest on the Moor itself lost the young to a fox, so it is interesting that the young in one of these forest nests have survived when I had seen a family of foxes so close.

At the first of the two harrier nests mentioned on 12th June we saw the cock drop prey before we went near. No sign of the hen at first, then as we collected one of the three big young from a ditch, we heard her shrill whickering and she came at us in a long low dive and thumped the back of my head with her foot. My old 'harrier hat', well holed with claw marks, is some protection. As she banked and turned to attack again, I tried to memorise flight attitudes and the marvellous chequered pattern of her underwings, always more exciting than in anticipation. Briefly she perched on top of a small pine and we could see a ring on one of her brilliant yellow legs. I can only guess that I ringed her as a chick in a neighbouring territory. We dare not stay long; a licence to examine harrier nests does not lessen one's sense of intrusion. There is so much to absorb quickly; the brood consists of two females and a male, the latter grey-eyed and thin-legged, much the most active and inclined to fly. Colours of legs and cere are already intensely chrome-yellow, their backs have a purplish bloom, short tails beautifully barred with light red. Mostly they sit back on flattened tarsi, panting and lolling their tongues, but as we approach one of the big females she raises herself in threatening attitude with wings half-spread, mantle and facial ruff inflated, part owlish, part catlike. Food has been taken to the nest but also to an extra feeding platform recently built a few feet away. As usual, when the young are big, larger prey has been brought, including an adult Red Grouse.

The second harrier nest was a different story. For an hour we watched from a commanding ride to make sure that we had pinpointed the nest site correctly. At last the cock, with foot down, sailed languidly above the trees and we heard the hen's mew-call as she rose for the pass. When we went to the nest she flew away and hung above the hill-top so she is probably the same timid hen which nested here last year. Though a little bolder, the cock, too, kept his distance. The nest had only two young, one of each sex, and near to fledging. Has there possibly been egg-eating by the hen as certainly happened at one nest lately? On the way up L. had spotted a party of Crossbills flying over and I was reminded of a harrier nest in 1966, where a Crossbill's head was among the prey remains. At this nest today we only found one small Willow Warbler chick, hardly fledged. This was a clean nest compared with some that I have seen.

12th July. A long day with A. at the lochs below the Seven-mile Ridge and up on to the tops, more memorable for the stimulating talk than for noteworthy birds. A. is an all-round naturalist and farmer of great experience—indeed he introduced me to this wonderful countryside more years ago than we care to remember. He is also a keen shooting man and on this subject we agree to hold differing views. Our hill climbing was punctuated by pauses and diversions, when A. lunges with his net after tipulid or bumble bee which he invariably identifies. On a green dripping ledge parsley fern is found, and higher up some fine clumps of starry saxifrage, but botanists would not compare these hills with some, for lack of limestone. After an hour on the tops my rucksack is weighted down with fragments of rock, some splashed with yellow lichens, others covered with woolly fringe moss, which I take home to add to the litter of 'raw material' in my studio. From the Ridge we looked down a steep escarpment of dark rock faces to a deep loch below. Even under this cloudless summer sky this can be a brooding place, awesome in wild weather. Suddenly a Peregrine cackled and then a pair with two strong flying young were all over the sky. No doubt these exhibitions which look like pure enjoyment are essential training for the young. At least it can be said that the time when pole traps were set for them here is past.

We recalled the July days, some years back, when we searched in vain for proof that a pair of Red-throated Divers were nesting at the loch. On a broiling day that summer I swam to the tiny islet where Otters had their home, complete with midden and a steep slide into the water. Might they have taken the young divers? I often wondered. In recent years Black-throats have been more successful. Today the loch was birdless except for a late family of Common Sandpipers on a little beach of sparkling sand and three moulting drake Goosanders.

17th July. At the Eagle wood the nest was empty, but we must have underestimated the eaglet's age on the June visit, as we found it eventually in a tree near the edge of the wood where it had probably flown with the help of yesterday's high wind. It seemed quite unable to take off and finally slumped heavily to the ground when we were just below it. It was so close that we clearly saw that it had a slightly deformed bill, with crossed mandibles. Would this be a serious handicap? Perhaps not, as so far it has obviously fed well. One can only speculate on an explanation for the deformity though nagging thoughts of toxic chemicals causing malformations crossed my mind. It stalked away through the tall bracken at surprising speed and jumped up on to the remains of an old dyke.

Once more we retired to the opposite slope to await events. The hill was quiet except for the occasional distant piping of a Golden Plover but the brackens were swarming with the small moorland birds, Whinchats, Wheatears, Meadow Pipits and Reed buntings. Perhaps these were the attraction for a young male Sparrow Hawk which came across the open moor. At last an adult Eagle—the cock from its dark tail—appeared carrying a kill and alighted on a dyke down below the trees. By this time the eaglet had ambled a long way down the hill and was quite close to where the adult landed. It made no attempt to approach the parent as it tore at the prey, but instead, as if under orders, began to move at a fast shambling walk back towards the wood, looking rather like a big, dark brown turkey. Was it simply that it knew that home was where food would be brought or had it responded to some signal from the old bird? It retired into the cover of the wood and we saw no more of it. It was not until half-an-hour later that the old cock eagle, having fed well, took the rest of the kill up to where the eaglet had disappeared.

31st July. After a fortnight away from home there were many signs of seasonal change as we drove up the valley. Curlews and Lapwings, which began to leave the uplands in June, are now gathered in flocks of many hundreds on fields and marshes. Redshanks had nearly all left by early July and most of the Oystercatchers are now away from inland haunts. Just at the time when holiday visitors are flocking into the district, most of the hill birds have begun their autumnal dispersal.

AUGUST

2nd August. The Nightjars in the forest above the Flow have always been baffling. Two years ago, in late July, L. and I spent several evenings watching a pair with a territory bordering a forest road. On bright nights the male was often visible churring from the top of the tall spruces, and from the female's behaviour, we were sure she had young quite near. She hovered around us and alighted on the road, jumping up and landing again just ahead of us, calling 'chook' in alarm. Then we would glimpse the male in low twisting flight among the trees and hear him call 'kee-wick' and crack his wings together. But this forest habitat is difficult country, especially in semi-darkness, and we could find nothing though we searched all the barer patches of ground where the nest would surely be. Most of the ground cover, of course, is extremely dense among the trees and does not appear to provide suitable nesting sites. Local information is that they always used to nest in the old hardwood, a remnant of which survives among the conifers and we still hear them near there. I have heard, too, that in another forest they deserted an old haunt when a ride was made into a road.

It should not yet be too late in the season so we set out for another try. But first we spent an hour, before dusk, by the Forest Loch. Lovely calm evening, ferocious midges, of course; yellow water-lilies flowering beyond the reeds. A party of Carrion Crows on the hill across the loch began to call and look restless—then L. spotted the fox which had alarmed them. It was quickly lost in high bracken but we picked it up again when a Curlew which had been quietly feeding leapt into flight, chuckling in alarm, and flew at it from fifty yards away. Swallows began to gather in the reeds and a duck Teal with seven quite small ducklings zig-zagged frantically as she started to cross an inlet just in front of us. Then seven Goosanders flew upstream, probably a duck with a fledged brood, going fishing in the river. At about ten-fifteen we moved up to the Nightjar zone. A Roebuck barked furiously. The churring began and, moving along forest roads, we heard at least four different birds. But all were deep into the trees and we had still not even seen one when darkness was almost complete. In total darkness we abandoned the Nightjars and followed the food calls of young Tawny Owls, finally catching one in the headlights of the car, and saw the adult come down with food. It could be three months since these young Tawnies were out of the nest and we found another brood still calling for food on the way home. A Barn Owl was hunting, and probably carrying prey to its young in one of the deserted cottages which provide many nesting sites in these forests. Earlier we had seen Curlews crossing the sunset sky and heard one burst of wonderful song, nostalgic of spring.

8th August. Last week I went to a distant grouse moor with two ladies to see a startling discovery they had made. It was a Crow Trap of the usual type but they had heard from a forestry worker that it was apparently being mis-used. On their first visit, a week ago, they were horrified to find three Kestrels and two or three thrushes, alive and struggling to get out, and of course they released them and left the trap open. They told me there were also a number of dead birds in it. So next day I went with them and examined the trap thoroughly. I counted nearly forty dead birds, mostly fairly fresh, a few long dead. There were at least twenty-five Mistle Thrushes and four Ring Ousels, all juveniles, two Cuckoos, a Blackbird and some remains of Red Grouse and Greyhen. According to workers in the neighbouring forest a keeper was often seen in the vicinity so it did not look as though this was a simple case of a trap being neglected. I contacted the R.S.P.B. immediately and the police have been informed. Now, today after the R.S.P.B. and the police have visited and photographed the trap and seen the keeper's employer, a Duke, I am told there is no case for a prosecution. Apparently the trap is standard $1\frac{1}{2}$ in mesh and there are nine others like it on the estate. The keeper will deny knowledge of protected birds being caught, but of course such traps must be visited daily. Further, if it had not been visited, how to account for no sign of dead birds of prey, when three Kestrels had got in very recently ? It is a murky incident and, in my view, should be publicised as the dangers of these traps, if only from neglect, are not recognised. At least the R.S.P.B. are taking up the question

of mesh size with the Ministry of Agriculture who are responsible for the speci-
fications. July and August are dangerous months for birds of prey in that area.
A few years ago we found pole traps set on another moor, where harriers have
bred, but could get no prosecution in spite of evidence that Kestrels had been
caught and this year there has been another pole trap reported.

13th August. At the Moor a Snipe rose from beside a young one hardly
able to fly. First broods were hatched in May so are these August chicks possibly from a third laying? I watched a Short-eared Owl with a vole today; it
landed on a road and ate it piecemeal, not swallowing it whole as I expected.
A Greyhen had a brood of half-grown chicks, one of which had been killed on
the road. The Blackcock are all in separate parties, six already at the orange
berries in a rowan tree and three more under waist-high bracken where they
had been feeding on grass seeds. Some have not yet grown their tail feathers
after the moult, and have barred brown feathers on their unglossy backs—the
'eclipse' plumage. Five Kestrels were hunting the slopes of the little ridge—
no doubt the family from the Lion's Head, which were reared in an old Raven's
nest. After sunset one was still hovering in the sky. The stone dykes seemed to
be alive with handsome Wheatears, in all stages of changing plumage. In
another week or so hardly one will be left on the hill ground. Tomorrow a
party of guns are coming for the grouse; there will be no driving to butts and
only a few brace will be shot, but if the grouse are as scarce as I think, the
shooters will say there are too many birds of prey and a passing harrier may not
be safe.

We took the car up a grassy-centred by-road and stopped beside a long
meadow, bright with creamy meadow sweet, red sorrel, deep purple marsh
thistle and the beautiful tall heads of lilac-blue sow-thistle. For two hours we
parked on the road—no other vehicle came—while I made sketches of the
roadside birds; no less than ten juvenile Whinchats, several Linnets, Bullfinches and Goldfinches and a family of Partridges, with Swallows and
House Martins continually skimming the tall vegetation. Here was the essence
of August, in the foreground the profusion of birds and flowers, the patchwork
fields and gleaming white farmhouse across the valley, and beyond, the hills
draped with the blue shadows of big slow clouds.

25th August. Along the road to the Dark Gairies, a third brood of young
Stonechats were not long fledged, a young Cuckoo flew over the heather and
Yellow Hammers were still feeding nestlings. I walked up a forest ride, lush
with a profusion of grasses. What beauty and variety of form they have, from
the feathery heads of Yorkshire Fog to the elegant tall sprays of Tufted Hair
Grass! Even the despised Purple Moor Grass is worth a close look at this

season—thanks to my botanist brother, I have admired through a lens the intense purple of its tiny spikelets of flowers.

Below the crags of Ben Gairy I disturbed a flock of fifty Mistle Thrushes which had probably been feeding on blaeberries. Ian, a forest ranger, lately shot a fox which had its stomach crammed with blaeberries, together with two naked young voles. A Roe doe, in red summer coat, and her three month old calf, were browsing some spruce which had grown like shrubs from earlier deer damage. It is Forestry Commission policy now to tolerate a permanent stock of Roe Deer but the ranger must shoot a percentage each season. From figures I have seen this may be something like half the estimated population.

Back in the valley I found a flock of seventy Goosanders on the big loch. These gatherings, of ducks and young birds, are at their peak now. The drakes, which left the incubating ducks in May, are nowhere to be found.

SEPTEMBER

1st September. I have been going over the season's cards for the B.T.O.'s Atlas of breeding birds and summarising success and failure of schedule 1 species for the R.S.P.B. Taking the region as a whole it has been a fairly average year. Human interference caused some failures among Eagles and Peregrines and the latter still appear to be affected by toxic chemicals, but less than in the early 1960s. More pairs of Hen Harriers were on the ground in the spring than were proved to nest and only about half of these reared young. Clutch size in harriers has been lower than in some earlier years, four being the maximum this year. In the past, the largest clutches I have seen, of seven and nine eggs, were in moorland nests and there seems to be a tendency for forest nests to have smaller clutches, though not necessarily lower fledging rates. Wet, cold spells in the spring may have reduced harrier success and certainly resulted in poor survival of young Short-eared Owls. There is no sign that weather affected Eagles, Peregrines or Merlins; of these three, the Merlins are the safest from human disturbance because they are less easily found.

Atlas cards, recording the species in each ten kilometre map square, show some interesting patterns of distribution. Redshanks, for instance, are common in the alluvial river valleys, almost absent from the peaty heartland of the hills. Leaving the higher ground in the forests unplanted is no help to Golden Plover as lack of grazing or burning allows the heather to grow too long for them and they have always been scarce on the bare high tops. We cannot find Twites nesting in these hills, although they have spread or been re-discovered on neighbouring moorland, near coasts, and might be making a come-back. The spread of the conifer plantations has been favourable to Blackgame but there are signs that they are being shot almost to extinction in some privately-owned forests—a small local nature reserve with excellent habitat for them cannot hold a population because nearly all are shot when they stray into adjacent forests where the shooting is let to irresponsible tenants from distant places. This kind of shooting undoubtedly affects other species too, including protected birds of prey. So far a quantitative study of breeding birds in sample areas of local forest has not been done but indications emerge from notes made by G., a young forestry worker, and ourselves. It is obvious that the retention or planting of more hardwood would enormously increase the variety of bird-species nesting, but there is little sign that this will be done on a significant scale. The large blocks of young conifers have a high breeding population of some small birds—Robin, Wren, Chaffinch, Willow and Grass-hopper Warbler, Meadow Pipit, Stonechat, Whinchat, Redpoll, Goldcrest and Coal Tit, with both Thrushes and Blackbird a little later. Many other species are much more patchily represented and may depend on factors such as proximity to hardwood trees and bushes or open ground at the woodland edges. Thus Bullfinches, Linnets, Yellow Hammers, Dunnocks, Tree Pipits, Chiffchaffs, Whitethroats and Garden Warblers are all more local. Even Whinchats are to some extent edge birds. Reed buntings have been notable colonisers and are increasing. G. has found that nests are often beside forest drains. There are notable absentees among tits and finches. Though neither Blue nor Great Tit can be expected, Long-tailed tits seem likely but I have not yet found them nesting in blocks of pure young conifer. Greenfinches are conspicuously missing, so apparently are Goldfinches, but Siskins have been welcome newcomers. Now in early September, there has already been an exodus of many of the small breeding birds from the young forests.

13th September. My daughter K. and I walked up to the little ridge on the Moor on this warm autumn afternoon. On the higher ground bracken was yellowing but the big patches were mostly still green. They looked blue-green against the sun, much lighter than the purple heather slopes. On the flat ground below the Lion's Head mats of deer grass smouldered orange-tawny, now the most brilliant colour on the moor. All detail was lost in the distant cloud-shadowed forests and on the deep indigo-blue mountains beyond. How silent the Moor is now, a silence that will hardly be broken, except by the voice of the shepherd, and the calls of Grouse, Crow or Raven, till the Curlews and Larks come back in March. Parties of Meadow Pipits passed above our heads, in high dancing flight. We disturbed one Skylark and other, single

Meadow Pipits. A young cock Reed bunting was still in a brackeny hollow where a pair nested. We found one of the dark-brown golden-banded fox-moth caterpillars in the heather. No doubt some grouse were quietly feeding somewhere on the hill but we neither heard nor saw any. As we came down a beautiful Short-eared Owl floated across our path. The orange-buff colour on its wings suggested a bird of the year.

24th September. Near the Dark Gairies I spent a fascinating ten minutes with a Peregrine. When I first saw it, it was silhouetted on a distant rock and might have been any large predator. It took off and flew rather high above me, not seeming to hurry. I followed it with binoculars and reckon it had flown three quarters of a mile when it suddenly stooped at a flight of small birds which I could only just see in the glasses. It missed and circled for a moment before flying steadily back to its original perch. I have no doubt it had been watching the sky and had seen the small birds before it took off.

It was a day of dripping grey skies with the moorland colours, now at their richest, glowing in this sombre light. It is on days like this that I am most conscious of the aesthetic loss as tight rows of conifers replace the tapestry of the moor almost all the way from the valley to the Dark Gairies. How extra-ordinary that planners can solemnly designate routes of great scenic value and say nothing when the character of the scenery is totally changed. Yet as I try to explain to foresters, I am *not* opposing all planting in these hills, but pleading for variety and the retention of sizeable enclaves of open moor.

The burn below the Little Gairy was swollen from recent rains. Above the sound of the water I heard the sweet rambling song of a Dipper. Like Robins, Dippers sing strongly in the autumn and winter months, when competition for territories is acute.

It is the rutting season for the Red Deer. One roaring stag was silhouetted against the grey sky on the brow of a forested hill, his hinds among the trees below.

27th September. Local bird-watchers tend to go mostly to the shore in autumn, looking for migrant waders and the first geese. Up here in the hills I have already seen a few Greylags coming south to join the resident birds in the valley. The presence of this small breeding population means that the first autumn arrival of Greylags no longer has quite the old impact on eye and ear and I shall feel more enthusiasm when the Greenland Whitefronts start to arrive, next month. In the forest below the Seven-mile Ridge a couple of Short-eared Owls and four Kestrels were working a slope for voles but these are low numbers compared with some years at this season. In 1960 a local vole plague brought memorable concentrations of predators; on 26th September

I counted twenty Kestrels, twelve Short-eared Owls, a Buzzard and three Hen Harriers over one block of young forest and the adjoining grassland. The grass was riddled with vole runs. The harriers, particularly, hunted the open grass-land and I saw one cock catch three voles, one of which he dropped and did not bother to retrieve. So, at times, harriers must be considerable vole-hunters. I have often wondered whether a local abundance of voles accounted for the exceptional clutch of nine eggs in a nest, one year, at the top of the High Glen; that pair had several pairs of Short-eared Owls as neighbours. Harriers are difficult to find just now so I went to watch one of the nesting territories at dusk to find out if any were still roosting there, but I saw nothing. Sometimes I have seen a family still using the nest for roosting for a few weeks after they could fly but they seem to disperse during August and the young at least may leave the district then.

MOUNTAIN AND MOORLAND

13 Willow scrub in an upland valley. *Photo : Jeffrey Watson.*

14 On a high ridge in winter. *Photo : Dr H. A. Lang.*

MOUNTAIN AND MOORLAND

15   Moorland and crag. *Photo : Dr H. A. Lang.*

16   Closed canopy spruce forest. *Photo : Jeffrey Watson.*

OCTOBER

1st October. This year the equinoctial gales from the west have missed us. Consequently there have been none of the storm-driven strays which sometimes turn up at this season. One year it was a Manx Shearwater which was found at the doors of a local power station and lived in our house for three weeks until strong enough to be launched from a sea cliff. Once, in early October, a wildfowler shot a young Arctic Skua in the valley and another year we added Velvet Scoter to the long list of ducks seen on the big loch. Grey Phalaropes, Gannets, various auks and even a Pomarine Skua have been found far inland in the district during this month. There is always the chance of trans-Atlantic migrants, like the Wilson's Phalarope which E. found in August. A year ago, in late September, L. saw a small Nightjar-like bird hawking over the river at dusk and was sure it was an American Nighthawk. It had broad white splashes on the wings but lacked the white tail spots of a male Nightjar. Exciting though these rarities are, there is enough of interest without them. It is time to begin the regular winter watches at the Hen Harrier roost again; a single cock came in on 17th September but we do not expect numbers before October. Why should this be so? Are the communal roosters all winter immigrants from other regions? This begins to look possible, but why do October counts at the roost generally show a majority of adult cocks? There is some evidence that ringtails, especially young birds, are more inclined to move south from the more northerly breeding grounds in autumn.

289

A fresh south-east wind was blowing as I approached the Flow tonight. Colours were rich at the forest border, whin and cross-leaved heath still in flower, tall bracken every shade of green, yellow and madder brown, yellowing larches above. A passing Carrion Crow was slow to spot me among the trees, but saved its alarm call until out of shot range. A party of Pied Wagtails bounded past on their way to a roost up-river. The big, early autumn gatherings of Pied Wagtails are past their peak; many have gone south already. Half an hour before sunset a Green Woodpecker called loudly from the forest behind me, where the old birches are. A party of Blackcock flew over the Flow and pitched for the night, surprisingly close to where the Hen Harriers roost. I did not expect many harriers at this date, certainly not as many as the five or six cocks and single ringtail which all arrived within twenty-five minutes, before sunset. The light was still good and for a few moments I had a breathtaking view of three cocks patrolling the river bank just opposite. Two of these had dark mantles, sharply contrasting with light grey wings—are these second year birds? If so, this could be their first time at this roost. I can only guess that newcomers find it by following birds with previous experience—perhaps it was significant that three cocks arrived almost in procession.

While I was down by the Flow I could hear two stags roaring on the hill behind me. Ian, the ranger, agrees with me that walking through a darkened forest with stags roaring among the trees, can be rather frightening. He has good reason to think so, having twice been charged by rutting stags, one of which he had to shoot when it showed no sign of stopping at fifty yards. Apparently it is more likely that the very large antlers on some local stags are due to good feeding during free range through the forests, than to ancestral inter-breeding with introduced Wapiti, as has been suggested.

8th October. Never has there been such an Indian summer as this. It has been so warm that I have been painting outdoors in my shirtsleeves. For three long days this week I have been working at the same spot, by a bridge where the river enters the main valley from the High Glen. I am totally absorbed by the richness and complexity of colour, tone and texture. It is a physical and emotional struggle, with much frustration and near despair, but has moments of extraordinary joy. I am so committed to my painting that I probably do not notice all the birds which come near me. At different times I am aware of a Dipper, a party of Mistle Thrushes, Skylarks moving down the valley, and the occasional hollow voice of a Raven in the sky. Once a flock of Longtailed Tits almost brushed my face as they flitted between the birch saplings along the river bank.

I returned home to find that J. had found another injured Barn Owl on the road. We already have two owl casualties in our charge, one Barn and one Tawny. Tonight they are all settled together in an unused basement room.

9th October. Common Gulls and Jackdaws were foraging among the grass on the high rocky knowes below the Seven-mile Ridge. Conditions are very dry now, difficult for ground feeding birds. Normally the big flocks of Common Gulls, which spend the day in the fields, start to flight back to the estuaries in the afternoons but they had made no move by late evening today.

15th October. There has been a resurgence of activity among the Red Grouse at the Moor. Cocks, showing quite prominent combs, are intermittently aggressive and noisy again, establishing autumn territories. A flock of twenty Redwings and several parties of Meadow Pipits crossed the Moor from the north-west. Are these all Icelandic migrants? Certainly it is sometimes noticeable at this time that Redwing movements coincide with arrivals of Wigeon and Whooper Swans, which pretty certainly come from Iceland. It is tempting, also, to suppose that the marked increase of Merlin sightings from early October is due to immigration of Icelandic birds; likely enough they would follow the pipits and Redwings. There was a Merlin, this afternoon, on the heather just below the Lion's Head, close to where a nest was found last year, so this could be a local bird. In a recent publication Leslie Brown writes that Merlins desert their upland breeding haunts for the winter but this is not entirely true in this region. Two years ago, at this date, a very beautiful young cock Merlin, which had been injured by a car, was brought to me; it was of the Icelandic race, by wing measurement. Late Wheatears are unusual on this hill-ground, though passage continues on the coasts not far away even into November. I looked along the forest boundary for a Great Grey Shrike without success. After two good winters for shrikes it looks as if this is a lean year for them here.

25th October. A long day's painting near the Flow gives an impression of the volume of bird movement into the district. Many Redwings seem to have moved on but the passage of Fieldfares is reaching a peak. I like to see them peel down from the sky and crowd on to the hawthorns for the crimson berries. Two small skeins of geese passed over high from the north-west. This is the usual flight line for arriving Greenland Whitefronts, though very often I do not see them until they have settled in the valley fields. It is six months to the day since I last heard their high-pitched 'wick-a-wack' calls. Some years a few arrive early in October but full numbers may not be in till the end of the month. One particular estate in the valley, where shooting is carefully controlled, has become the Mecca for the geese. At the end of the month I expect to find up to a thousand Greylags and four hundred White-fronts there, with Whooper Swans and a fine mixture of duck in the marshy lagoons.

As I packed up for the day the rising wind was fluttering the last of the small yellow birch leaves. A gale is forecast and tomorrow the leaves will lie in a wrack against the river bank.

28th October. Early this morning when I drove past a frosty field alongside a forest, five Blackcock were parading and challenging in full display posture. In the soft light the 'white' under tail coverts reflect blue and yellow from the ground below. Their wings almost meet beneath the fanned and stiffened tail feathers. The throaty crowing is made simultaneously with a ridiculous little jump. They look pompous and surprisingly athletic in turn as they prowl and scuffle together. No Greyhens attend these autumn leks and some of the cocks soon lowered their tails and began to feed.

NOVEMBER

2nd November. On my way to an evening lecture I stopped below Ben Gairy. Ten minutes before sunset the pair of Eagles came in to roost near the eyrie. As they approached the cliff the cock made a brief display flight, rising and falling steeply. I hear that the young Eagle from the tree-nest has not yet left its home-ground—three eagles have been about there through the autumn. I should think it may leave soon as one or two eagles in white-tailed plumage have recently been reported, distant from known nesting territories. The breeding pairs roost near the eyries through the winter. One pair seem to spend most of their time within a radius of five miles of the eyrie, but they may sometimes hunt much further afield. We have been trying to learn more about how widely the harriers travel from their roosts to hunt, but this too is a difficult problem. One ringtail, last winter, regularly hunted an area between six and seven miles from the Flow, and its evening flight line was observed. But this year, so far, no harrier has been seen hunting this territory. They certainly appear to have distinct winter hunting ranges, though these may change from year to year and even during one season. They are particularly attracted by fields of kale or turnips where flocks of finches and other small birds may gather. Several times recently I have watched a cock quartering an area of tall marsh grasses. It is not easy to follow a hunting harrier for very long; superficially its flight may look lazy but it can move with astonishing speed and purpose, flying so low that it may be lost behind the minimum of cover. Then it may drop to the ground and remain there for long periods—in their study of the American race of the Hen Harrier the Craigheads found that five-sixths of the time was spent on the ground in bad weather.

6th November. The hedgerows down the valley are full of Blackbirds. The big immigrations of Continental birds are always apparent about this time, slightly later than the first big waves of Fieldfares. A high proportion are first year males, with dark bills and sooty plumage. A friend tried to convince me she had seen a 'melanistic' Blackbird but of course it would be one of these young males. A keeper tells me that few Woodcock have been seen during local Pheasant shoots so far. It was on this date some years ago that I was watching Blackbirds prior to roosting when a Woodcock flopped on to the grass among them. Apparently disconcerted at finding itself surrounded by jostling Blackbirds it adopted an aggressive posture with its tail fanned, displaying the startling white tips on the under side, and twice made little leaps in the air. This performance scared off the Blackbirds and it began to prod with its bill in the leaf litter. The light was failing but I continued to watch it feeding and preening and I could hear the crack of its carpal joints when it raised its wings above its body. From looking squat and almost hedgehog-like it suddenly stood surprisingly tall.

15th November. Floods have brought new forms to the landscape. A stone dyke, unnoticed before against the level green of the turnip tops, cuts a curious pattern, like a derelict jetty, in the still water. Its jagged outline contrasts with the rounded shapes of sleeping Mallard. The November passage of Teal is at its peak; all the way down the valley the flood-pools are crowded with little flocks and there have never been so many Shoveler as this year. After the storm came frost; in the woods the smooth trunks of stripped beeches stand blue-grey on a mat of wet purple leaves, where Woodpigeons are searching for mast. In these surroundings they are beautiful birds, rose-grey, like old china in a cool room. Over towards the Flow wreaths of white mist lie across the forested hill; lower down the fringe of birches is bare and dark as wine, but many oaks still hold their tight canopies of brown and yellow leaves. I watched a flock of Redpolls in the birches. Some were picking at the lichen-covered branches, presumably finding seeds that had been ensnared. It is some years since I identified Mealy Redpolls, but this must be the time to look for them. Blackcock, awkward but acrobatic, were among the high twigs, nipping the buds already. Now and again a female Capercaillie has been reported among them here, likely enough correctly, as the females are the pathfinders beyond their established range and, of course, the species has now been introduced by the Forestry Commission to northern England.

A Great Spotted Woodpecker bounded over the forest, heading for another, distant, enclave of hardwood. Many Chaffinches and Robins are tame at picnic spots within the forests and immediately fly to a parked car. I wonder how many of the large autumn population of forest Robins survive the winter. G., working all winter among the trees, has found that most disappear when weather becomes severe. Jays, which were so difficult to see in summer, have lately become conspicuous and noisy again. They have become more wide-spread than Magpies in the older conifer forests, though both are still increas-ing. A. considers that local Jays originated from Scandinavian immigrants. Most are distinctly greyer than typical English Jays, but two which were brought to me alive by a keeper were dissimilar, one redder than the other. My chief memory of these birds, which we kept for a few days before releasing, is their extraordinary power of mimicry. They could imitate perfectly the 'tew-ick' and hoot of a Tawny Owl and the alarm calls of Kestrel, Carrion Crow and Blackbird; an object lesson on the unreliability of call-notes as iden-tification. Mimicry occurred when they were alarmed by the approach of a man. I also remember minutiae of colour, like the rich burnt sienna notches on their inner secondaries and the brown outer ring to the pale blue iris.

26th November. Visits to the harrier roost to observe morning departure are sometimes not too successful, due to poor visibility. At least it seems certain that the totals counted in the evenings are not increased by birds arriving after dark. They start to leave the roost about half an hour before sunrise, and most slip away quickly. They may circle high above the roost before setting off, but in this terrain we cannot follow them out to distant hunting areas. After a mild, drizzly day with fresh south-west wind prospects seemed good tonight. We were not disappointed but the build-up in numbers of ringtails has been much less than in past Novembers. Last year they began to exceed the grey males from early November onwards; on 16th November, for example, we counted twelve ringtails to four grey males. This year the latter have continued in the majority until now. Tonight, for the first time, ringtails just came into the lead, with nine to eight. A total of seventeen birds is the highest this autumn but well below some November peaks in the past. I suppose it is possible that a generally poor breeding season could explain this. It is impos-sible to distinguish first year from adult females at the roost but first year males can be recognised, as small ringtails, if they come at all near or can be compared with a female for size. During a previous November a ringtail, almost certainly carrying a wing-tag, was seen several times and shortly after-wards a report came in that a young male wing-tagged in north-east Scotland, had been found dead not very far away. This was the first proof of autumn immigration into the district. Only one young male was seen tonight. It was a fascinating watch, in spite of almost continuous rain and early murkiness. The birds mostly came battling up the valley in the teeth of the wind and rain, as if drawn by a magnet to the roosting ground. Several times males which had settled were driven out by larger females and forced to look for a new spot. This winter a preference is being shown for reed beds over the tall grass and rushes most favoured in the past. A daytime visit to examine the roosting forms and collect pellets is overdue but the ground may be too waterlogged at present. The last analysis of winter prey remains showed that the harriers were feeding mostly on small birds, from Starling size downwards.

30th November. On a bitter grey day with snow on all the high ground I saw a small flock of Snow Buntings on a low hill near the entrance to the High Glen. A few were reported on the tops in October. A single Waxwing has been seen in the village and the first small flock not far away.

DECEMBER

6th December. A pair of Red Grouse looked plum-red on umber-green heather on the Moor. A keeper tells me that on the best grouse moor in the neighbouring county there has been a catastrophic fall in numbers after some seasons of great abundance. I do not know the reasons but it sounds as if the heather may have suffered from frosting in winter and early spring and the grouse stock may have risen too high for the habitat anyway. The keeper (who is more concerned with Pheasants) agreed with me that the situation showed that there was no case for destroying birds of prey on that moor, as I know is done. Here on the Moor, there has been some talk about the scarcity of grouse and it could be that a full study of predators and prey would raise awkward questions. But I wonder if, in any event, there can be much future for grouse on small moors like this, where the heather is not managed for them and the growth of the surrounding forests is eliminating the heather on neighbouring land? All the same, with the growing pressures in favour of sport in the countryside the future for birds of prey in any wild country looks increasingly uncertain. Most of my birdwatching friends only come here on flying visits and it is left to the resident to ask the difficult questions and risk being branded a crank. I have heard startling figures of birds of prey 'bagged' by some shooters. But confrontations on this issue may achieve nothing; perhaps the best answer to a harrier-hater is to try and persuade him to come with you

297

and experience the delight of watching a pair in spring—I know one well-known ornithologist who has had some success with this method.

Up on the little ridge a pair of Great Black-backed Gulls stood sentinel. They are often the most conspicuous birds in sheep country in winter and the herd dislikes them as much as, or more than the Crows, especially at lambing time. One year Jimmy had some trouble with Ravens but he regards this as most unusual and seems to bear no malice towards them in general. I saw the pair circling over the little group of pines where they have had a tree-nest in most years.

From the little ridge it was a wild and desolate prospect today and I soon turned to come down, as heavy rain began to fall from banks of dark ragged cloud, intensifying the bracken colours.

10th December. With J. and L. to the Flow; rain cleared as we arrived. We decided to look for a new observation point, opposite the reed beds where most of the harriers have been roosting lately. We had difficulty in finding a position combining a good outlook with some concealment. When two cocks and a ringtail came almost directly over us I wondered if we might have a repeat of an earlier December evening when a ringtail dived down at me, whickering, and was joined by five harriers from the roost, circling close overhead. I always suspected that the first, deliberate attack was made by a bird which recognised me from previous summer visits to a nest. I shall be told this is far-fetched but Douglas Weir is positive that the Buzzards he has studied recognise him individually. But, although some birds saw us tonight, none showed any aggression. The evening became memorable, partly for the close approach of the birds and some wonderful flight mastery in the strong wind, but also for the brilliant after sunset light which illuminated them against an inky thunder cloud. A score of nine grey males, several with very dark mantles, and nine ringtails, continues to show an unusually high proportion of cocks for December.

24th December. Every day now there are one or two Pied Wagtails on the village street. They run about with little regard for passers-by and are evidently efficient at avoiding cats and cars. The great majority of the large summer population leave in autumn, yet there are always a few which stay, or come into the area, for the winter. At this time they are often in attendance on man or his works, around farm middens, in towns and villages, at rubbish dumps. I was surprised to hear that several were recently following forestry ploughing far up in the hills, until the snow came. Grey Wagtails have entirely deserted the hill ground for the winter but odd ones can still be found in the valley. A small difference in height above sea level seems to be important to Song Thrushes; although our garden is barely three hundred feet up we have had no Song Thrushes this month. But many of the breeding Blackbirds never go away; one old cock, with a growth on one foot, can be seen any day defending his nesting territory. Lately a Sparrow Hawk has been killing some garden Blackbirds but the old cock survives in spite of his disability. I believe it is the experienced bird with an established territory, not necessarily the fittest, which is most likely to escape predation.

This afternoon we watched the harrier roost again. We expected a low total, as conditions were windless with frost still whitening the grasses. In fact we did not see a single harrier. Visibility was poor, so we might have missed one or two. It is still uncertain why a change in weather makes so much difference to the numbers using this roost, but it may be significant that more have been seen down near the coast in the recent wintry weather and it would be a long flight back from there to the roost. There could also be difficulties in finding the way in misty weather. This certainly sometimes confuses the local geese, which have much shorter flights to their roosts.

30th December. A thaw is forecast after a week of hard frosts with light snow lying in the valley. There is no wind, a steady light and the hint of milder air, while the river is still fringed with ice and snow—this combination of conditions, which may only last a day, encouraged me to paint outdoors. I went to a quiet backwater, hidden from a nearby road by tall reeds, not far from the Flow. I might even see Bearded Tits here, they have become so farflung in their winter wanderings. A dozen Siskins were in the alders at the roadside, five Goldfinches below on the hardheads, where they still find seeds even in mid-winter. I was rash enough to try painting into the sun which, fortunately for my purpose, remained only a glimmer through mist. A subject to tempt the young Monet! Beneath a freckling of snow there was an underlying warmth in the dark red birches and yellow ochre reed stems, all sombrely reflected in scarcely moving water. I hurried over tiresome preliminaries, righting the easel as a leg plunged through crunching snow, fumbling with paint tubes, and set to work standing up to avoid painting too cautiously and also to keep the circulation going. After an hour despair had miraculously lifted and I thought the tones in my picture began to make sense. As always, on these winter days, the sounds of a shoot could be heard, mercifully distant. A keeper, passing on the road, greeted me with a friendly, 'this weather suit you?' In a burst of confidence I assured him it couldn't be better. No time to stop and talk birds to him now. Silence again as the keeper's crisp footsteps died away.

A pair of Goldeneye swished down on the water in front of me but rose again at once on taut, musical wings. I glanced up at a chevron of birds flying high and fast to the west—a flock of sixty Lapwings escaping from the snowbound countryside. Light began to fail, flattering my work. Stamping about, drinking thermos coffee, I heard, then saw the White-fronts crossing the forest in two long ragged skeins. Dusk and the ice on my palette were defeating me when I heard the soft bugling of Whooper Swans approaching. It seemed a long time before the little flock came out of the gloom and pitched at the edge of the reeds, still bugling nervously with their tall necks stiffly erect. As I stumbled back through a morass of partly frozen tussocks I put up a Jack Snipe and heard the shrill tittering of Little Grebes from the reeds behind me. Driving home I passed a Barn Owl on a roadside fence post.

31st December. Over Christmas and New Year there are many visiting bird watchers in the district. Twenty years ago it was almost an unknown country and it was an event when a party called at our house. I know that I am fortunate to live in this countryside and enjoy it through all the changing seasons. As its popularity grows, with all the claims of conflicting leisure interests, the need for planned conservation, of both landscape and wildlife, becomes imperative. In the New Year I shall be at a meeting to discuss some immediate threats, like high speed power boats on bird-rich lochs. Then, through the Scottish Wildlife Trust, I am involved in putting the case for preserving the Moor as an enclave of open country. In many and varied ways much of what I have described in this diary is endangered. Only last month a new forest road was begun below the Peregrine eyrie which we successfully guarded last summer. Other nesting sites are at risk from increased disturbance and accessibility as the hills cease to be a remote area. Yet, this region continues, rather precariously, to combine a rich and varied wildlife with unusual landscape quality.

ISLAND

by MALCOLM WRIGHT

CALF of MAN

Field Boundaries —————
Tracks ·········
Heathland ⊣-⊣-⊣-⊣

0   10   20   30   40 Chains
½ MILE.

N

ISLE of MAN

KITTERLAND

Little Sound

THOUSLA
The Harbour Sound
Grant's Harbour

COW HARBOUR
KIONE BEG

Fold Point

The Clets

James House ■ 144

The Heath

226

360

KIONE ROAUYA.

KIONE NY HALBEY

Giau Yiain

GIBDALE POINT
GIBDALE BAY

OBSERVATORY

Baie Fine

326

270

The Glim

AMULTY

421

Upper Lighthouse
New Lighthouse
Lower Lighthouse
Smithy

231

Baie'n Ooig

Crag Veanagh

Clet Elby

Culbery

Giau Lang

Millpond
Dub

Mill Giau

The Leodan

Rarick

South Harbour

THE BURROO

The Puddle
Manusen Rocks

CAIGHER POINT

THE STACK

CHICKENS ROCK & LIGHT
c. 1275 yards.

JANUARY

The Calf of Man is an island of 616 acres which lies off the south-west tip of the Isle of Man. It is separated from the main island by a Sound less than half-a-mile in width through which runs a strong tidal current. On the far side of the Sound lies an islet, Kitterland, which reduces the main channel of water at one point to less than 300 yards. From Cow Harbour at the northern end to Caigher Point in the south-west is one-and-a-half miles, and the island is a mile across at the widest point. The whole of the coast is rocky, with cliffs which in places rise to 300 feet, particularly along the west coast from Culbery to Gibbdale Bay, and on the east coast near Kione Roauyr. These steep slate cliffs provide an abundance of nesting sites for seabirds and eleven different species breed on them.

The island is virtually split into two by a shallow valley, known as the Glen, which begins above Gibbdale and slopes gently down to the lower southern end of the island. Most of the fields of the old farm system lie within the Glen. At the head of the valley, sheltered on three sides by the surrounding hillsides, lies the old farmhouse with its large garden which is now used as the head-quarters of the bird observatory. To the east of the Glen lies a large area thickly covered with heather, with some smaller patches of bracken, known as the Heath. The west side of the island is a mixture of heather, bracken and rough grassland. No trees grow native on the island but a few sycamores, ash and sitka spruce have been planted around the old farmhouse together with some bushes of wild fuchsia and elder. Because of these trees and its sheltered position this area holds a concentration of birds and many of the migrants come into the trees and bushes to rest and feed. A stream runs through the middle of the Glen into the millpond, which was artificially created to supply a water mill which once operated nearby. Just below the pond is a thick willow withy, which was probably planted to provide canes for lobster pots and basketry but which now provides excellent food and shelter for migrant warblers and other species. At the westernmost point of the island, 300 feet above sea level, stand two old disused lighthouses and a new modern lighthouse, completed as recently as 1968, which has a complement of three keepers.

After passing through a succession of private owners the island was presented in 1937 to the National Trust for England and Wales. It has enjoyed nature reserve status since that date and on the formation in 1951 of the Manx National Trust was leased to that body. Farming ceased in 1958 and in the following year a bird observatory was established on the island with a resident warden and assistant and accommodation for visitors interested in natural history. Many of these help with the work of ringing, observation and recording.

From the ornithological viewpoint the island has two principal attractions. The first of these is the birds which breed on it and especially the seabirds. About 4,000 pairs of birds of approximately forty different species nest each year. Secondly because of its geographical position it is an ideal place for the study of migration. To any bird approaching from the south over the sea it is the natural spot to make a landfall. Alternatively any bird moving from the north or east is likely to follow the coastlines of the Isle of Man and this will almost inevitably lead it onto the Calf.

Winter is the quietest time on the island. Many of the birds which breed are hundreds if not thousands of miles away. Young Kittiwakes ringed in the nest here have been recovered in winter in south Greenland; Meadow Pipits in Spain and Portugal, and the Swallows which breed in the old barns will have reached South Africa. Except in the event of a spell of unusually cold weather there is little movement through the island. Along the high cliffs the Fulmars are continually inspecting suitable nesting ledges, gliding effortlessly backwards and forwards for hours on end every day. Shags are numerous, both offshore and on the rocks near their breeding colonies and on a mild sunny day even in January a few will be starting to collect the first sticks for their nests. The Herring Gulls are often largely absent during the day in winter but late each afternoon several thousand come in to roost on the rocks and cliffs. The Partridges had an excellent breeding season last summer and there are several large coveys in different parts of the island. Woodcock can habitually be flushed from the same spots; around the withy, the back garden of the observatory and the marshy hollow below Janes House (an old farmer's or fisherman's cottage, now in ruins). One pair of Ravens are resident,

ISLAND 17 The Glen and the Observatory, Calf of Man—hills of the Isle of Man in the background.

ISLAND   18   Looking seawards, Calf of Man.

breeding each year on the mighty cliffs of Baie 'n Ooig and they will very soon
be repairing their often used nest. The crows of the Isle of Man are Hooded,
known locally as Greybacks and a few pairs of Magpies are also resident.

Of the smaller birds the commonest at this season is the Wren. There is
nowhere on the island where this hardy little creature is not to be found; in the
thick bramble bushes in the Glen; in the extensive tracts of heather or even
in the exposed cliffs. The Song Thrush seldom nests and the few which can be
seen are wintering birds, but the Blackbirds are resident, although only in
small numbers. One of the most characteristic small birds of the island is
the Stonechat and although part of the British population does move south
for the winter several pairs remain on the island throughout. A few pairs of
Robin breed but they are more numerous from autumn through to spring
and like the Wrens are often found in the rocky coastal regions. Rock Pipits
are numerous around the coast and while a few pairs of Dunnocks can be seen
in this region, too, most of them prefer the shelter of the bramble bushes
alongside the stone walls of the old farmland. All the Linnets which breed
have moved south for the winter but their place has been taken by their
northern relative, the Twite, and flocks of up to fifty of these attractive little
finches can be seen along the cliff-tops.

FEBRUARY

The seaweed covered rocks around the foot of the Burroo are a favourite haunt of Purple Sandpipers and a careful inspection at any time between October and April will usually reveal a few of these attractive little waders. The place to look is right at the waters edge, for they habitually seek their food in the zone between high and low water marks. Not even a heavy swell drives them to seek a more sheltered spot and they will run down the rocks after one receding wave to snatch some morsel of food then half-fly; half-run, back in front of the next wave, frequently getting heavily splashed in the process. They are very easily overlooked as their dark heather-brown plumage gives them a very effective camouflage, whether they are on the blackish rocks or the brown seaweed. Turnstones are also quite usual on the rocks around the coast and can be met with in parties of up to forty or more. A few Purple Sandpipers often attach themselves to these flocks and the two species regularly associate when feeding.

Most ornithologists have a favourite bird. Mine unreservedly is the Chough, to me the most delightful of all the many interesting birds which inhabit the island. This bird has personality in abundance and I never tire of watching them for they are always up to something interesting, amusing or unusual. On the ground they are often somewhat comical in appearance, swaggering around with a side to side walk but once in the air they are masters and have a graceful and elegant flight which few birds can equal. In character Choughs are very different from the other corvids. The Ravens, Hoodies, Magpies and others are cunning rogues, living on their wits and often bullying and butchering other birds, but the Chough depends solely on insects for its food and is as beneficial as it is gentle.

It is a rare event to see a solitary Chough on the island for they pair for life and keep very close together at all times. Almost invariably if one Chough is seen its companion will be found in close proximity. Frequently they flock, for the Chough is a very sociable bird. These gatherings seem to take place whenever two or more pairs happen to meet. Then the loud 'kee-aa' calls will echo across the Glen and the birds will fly towards each other, calling continually. Other pairs will hear the commotion and come flying in from Caigher Point or the east coast and soon eight or ten will be flying round in a compact bunch, all calling excitedly. Choughs from the mainland cliffs often visit the island so these flocks sometimes reach twenty or more. For a while the group will soar expertly in tight circles on outstretched wings and tail; then they will set off in some direction with an especially buoyant manner of flight, closing their wings and dropping ten feet like a stone before rising again almost as far on the momentum gained from the fall, so that they progress in a series of leaps and bounds. Suddenly the group will quietly break up and the pairs disperse to their respective corners of the island. Half an hour later, given a suitable opportunity, they will be back together again in another excited flock.

It is a great pity that this fine bird is now so scarce in many parts of Britain. Small numbers breed in Wales and a few of the islands of south-west Scotland but only in parts of southern Ireland is it reasonably common. Not a single pair remain in England and the total population of the Isle of Man probably does not exceed thirty breeding pairs so its hold here must be considered precarious. As recently as 1959 about ten pairs bred annually on the Calf of Man but this figure fell steadily in the early 1960s and the population now seems to have stabilised at some four or five pairs. The reason for the Chough's decline is not understood at all. Two theories often put forward are excessive disturbance by man and increased competition from Jackdaws, but neither of these will explain the recent decrease on the island. Because of the isolation of the place disturbance is non-existent and the Jackdaw, although common on the mainland, does not breed here and is only an occasional visitor. More likely the reason is a subtle one and is possibly bound up with the birds insectivorous diet. It is a very large bird to be entirely dependant on insects and does not seem to possess the adaptability of the other corvids which would enable it to cope with a changing environment.

The island enjoys a mild but windy winter climate. Lying snow is quite unusual and because of the warming influence of the sea the ground is very rarely frozen for any length of time. This year, winter has been mild throughout until a sudden cold snap set in on the 18th. A bitter cold north-east wind was blowing from first light and during the morning thick grey clouds, heavy with snow, built up on the landward side of the island. Soon after 10.00 hours I heard a distant squeaky 'pee-wit' and looked up to see a line of over 30 Lapwings high over the Glen flying steadily south-west. It was the first of many such flocks which passed over the island in the next few hours. Small flocks of Golden Plover also began to appear and raced across the island, more often heard than seen, and twice groups of Curlew called as they flew west. In mid-afternoon the snow began to fall, a swirling blizzard of white which made further observations impossible.

The next morning the sky had cleared but several inches of snow lay thickly on the ground. Birds were everywhere. Hundreds of Song Thrushes, Redwings, Fieldfares and Starlings had appeared overnight and a cloud of Greenfinches, Chaffinches and Bramblings flew up from the back garden. I cleared a few patches of snow and scattered some seed, and very soon dozens of birds were congregated at each spot. A few Lapwings and Golden Plovers were still flying west overhead and a flock of 80 Wood Pigeons followed in their tracks. A bewildered Woodcock, deprived by the blanket of snow of his camouflage, was walking along the path in the Glen and a dozen Snipe rose from the margins of the frozen millpond. It is when conditions of this type last for several days that the birds are unable to obtain food and severe mortality occurs. Fortunately this was not to be one of those occasions and a rapid thaw set in the next day before much damage had been caused. The extra birds dispersed as quickly as they had arrived.

MARCH

The wind became easterly at the start of the month and with the barometer rising the first sign of thrushes returning from their winter quarters came on the 5th with the arrival of about 30 Song Thrushes and 40 Blackbirds. Recoveries of Blackbirds ringed on the island have revealed that they are heading for breeding grounds in Scandinavia or north Germany after wintering in Ireland. A few are also British bred birds which have moved west. The Song Thrushes' destination is the northern half of Britain. The migrant Blackbirds are easily picked out as they are much more conspicuous than the residents, flying noisily away when approached, very often for a considerable distance, while the local Blackbirds skulk quietly into the bottom of the nearest bracken or bramble patch.

The 15th was a dry, mild day. I walked around Caigher Point and in an hour over 40 Gannets passed offshore, all adults and most of them heading strongly away north-west. Off the Point was a raft of nearly 50 Guillemots drifting slowly past in the current and looking resplendent in their chocolate brown and white breeding dress. Further round, in the Puddle, 24 Razorbills were bunched on the water but neither species had ventured ashore onto the cliffs. It is in fact extremely improbable that any of these birds had been on land anywhere since they left the breeding colonies last July. The auks are very tentative in their return to the breeding cliffs and sit around on the sea for days before they first come ashore. Then there is a period of some weeks

during which they come ashore at times but are equally liable to disappear
out to sea again, especially when the weather turns stormy. It is only when
egg laying begins in May that their attachment to the land becomes firm.

Most of the next week was cold, windy and wet but a break in the weather
on the 21st quickly set the migrants moving. Eighty Fieldfares, a similar number of Song Thrushes, 40 Redwings, 70 Blackbirds and a dozen Snipe were
all on the island at dawn. The Meadow Pipits have suddenly returned in force
too and at least 200 passed through during the morning while many more
were feeding among the long grass and sedge in the Glen and a few others
were singing. Undoubtedly some of the Meadow Pipits which breed have
been arriving for the past two or three weeks but they have passed almost
unnoticed during the gales and cold. The next morning the wind had veered
south-east and the first summer visitor duly appeared, a Chiffchaff feeding
quietly along the edge of the withy. Among the rocks in the Rarick was a
Black Redstart, a dull grey bird until it flew when it came to life with a flash of
its bright orange-red tail. The habitat preference of this bird is interesting;
each spring and autumn one or two appear and they are always found either
along the rocky shore or around the buildings and garden of the observatory
or the lighthouse, often staying in the same spot for several days. The first
three Wheatears were seen on the 23rd, several days later than their average
arrival date and an early Whimbrel was also found. A Yellowhammer which
flew along the fuschia hedge in front of the observatory is a rarely recorded
visitor. Although they breed commonly just the other side of the Sound and
within sight of the Calf no more than two or three are seen each year. Why
this should be so is a mystery as there would seem to be ample food and cover
for a few pairs to breed, but perhaps the island is too exposed for their liking.

At last the island was beginning to warm up and the next two days were
calm and sunny and it seemed that spring had come at last. Along the cliffs
more Wheatears were to be found, flicking away low across the heather when
approached. Meadow Pipits, the most numerous small breeding bird, were
singing everywhere, hoisting themselves into the air from tufts of heather and
the dry stone dykes. A dozen Chiffchaffs searched the few bushes and trees for
insects and occasionally one burst hesitantly into song. Two Peregrines
drifted slowly north above the west coast cliffs, clearly a pair as the higher
bird was appreciably larger than its companion. Were these just passing or
would they breed somewhere on the cliffs of Man? Until the population
crashed in the 1950s several pairs bred annually but it is now more than
fifteen years since breeding was proved in the Isle of Man. The Calf itself was
well known as an eyrie from the middle ages and they probably bred until
about twenty years ago. The crash is thought to have been caused by con-
tamination from persistent toxic chemicals, particularly of pesticides used

in farming, which cause eggshell thinning and in other ways interrupt the birds breeding cycle. The use of the worst of these is now restricted although unfortunately they have not been banned altogether. In the early sixties very few Peregrines were even seen on the island but the number of sightings has been rising steadily over the last few years and gives hope that it will not be very long before they breed here once more.

The Choughs are busy now rebuilding their nests. They are very traditional in their choice of nest site and very often the same nest is used year after year, a few sticks being added and a fresh lining. Two pairs of Choughs have been making frequent visits to the field where the Loghtan sheep are penned to collect scraps of wool rubbed off onto the fences and rocks by the sheep. I watched them strutting round the field, always keeping in close company and by the end of ten minutes they presented a highly comical sight, looking for all the world as if they sported large 'R.A.F. type' moustaches. Then they took off for the cliffs, calling all the way in spite of the wads of wool they carried. They headed directly for their chosen sites and it was a simple matter to discover that the cliffs of Gibbdale and Caigher Point would have Choughs as tenants again this year. Recently another pair were seen carrying wool to an often used site in the cave in the Eye of the Burroo.

Later that day I heard a lot of croaking and rushing outside found no less than 18 Ravens high over the Glen. The resident pair (who now have eggs in a nest at Baie 'n Ooig) were most upset and flew round the party continually calling and demonstrating that the island was occupied. The intruders, much to the annoyance of the residents, landed near South Harbour and spent an hour there before straggling back high over the Sound in several groups. One of the pair, probably the male, followed behind to see them off. Most Ravens are deeply involved in breeding by now so these birds were almost certain to have been non-breeders wandering about.

By the 29th the wind had moved round to a fresh, cold north-westerly. Small bird movement had come to a halt but 48 Lesser Black-backed Gulls were standing among the heather and bracken in the south-east of the island where they nest and had come in since yesterday. About fifty pairs nest in this area each year. At the main Shag colony near Kione ny Halbey over sixty birds were lined up on the rocks on the seaward side and as many again drifted on the sea nearby. They begin breeding early and already a few nests contained eggs, and many other nests were freshly completed. In one year, after a very mild winter, several nests with small young were found in late March which meant that nest building must have begun in earnest in early February.

Further up the coast on the Cletts, a series of low offshore rocks, twenty-one Atlantic Grey Seals were hauled out in the weak afternoon sun. They presented an astonishing range of sizes and colours, from glossy black old bulls seven feet in length to creamy-white young cows of under five feet. They are always present, though in variable numbers, in the bays and inlets around the island and must surely breed in the vicinity although their actual breeding caves have not yet been found. Not many other mammals inhabit the island. Rabbits were introduced long ago and can be very numerous but their numbers have been affected in recent years by myxomatosis which spread to the island. Pygmy Shrews are quite common, Long-tailed Fieldmice scarce and Brown Rats unfortunately reached the island after a shipwreck nearly two hundred years ago. Bats are sometimes seen flying around the farm at dusk.

APRIL

There is little sign of spring so far this month with a cold north-westerly wind keeping activity to a minimum. However overnight on the 4th it backed to a light breeze from the west and as soon as I stepped outside that morning I heard a loud 'clack-clack-clack' and a fine male Ring Ouzel flew out of the top of the sitka. Later it was joined by a second bird and they spent the day on the heather covered banks around the head of the Glen. A solitary Grey Wagtail was feeding along the edge of the millpond and Meadow Pipits were quite numerous with little parties of migrants rising up from the old pasture fields and flying off towards the Sound. Near South Harbour was a group of freshly arrived Wheatears, 14 in all and only one female among them. Most of the males were young birds from last year's nests, easily picked out from the older males by their much browner wings. The variation in the colour of their breasts was quite striking, grading from a pale buffish-cream to one bird with rich orange-red underparts.

The Puddle appeared empty apart from the usual Herring Gulls but the tide was low and a careful look round the exposed rocks revealed several species of wader. On the corner of the Rarick 30 Turnstones accompanied by four Purple Sandpipers were busily poking about on the weed. Two Whimbrels rose and flew off north-west, their clear piping whistles still audible when the birds were distant specks. Further round, 16 Oystercatchers were standing motionless, shoulders hunched, several of them on one leg and all facing directly into the wind. A couple of Redshank bobbed nervously up

313

and down and six Curlew paced sedately among the rock pools. Apart from several small pebble beaches the entire coast of the island is rocky so the sand- and mud-loving waders such as Dunlin, Knot and Ringed Plover are not often seen.

Later I watched one of those fierce little falcons, a Merlin, thread its way swiftly down the east coast, gliding along just above the cliff. Suddenly it banked inland and flew directly at a Meadow Pipit flying up the coast. The pipit saw the danger just in time, dipped in flight and the Merlin missed by inches. Turning rapidly it set off after its prey and after fifty yards was only a foot or two from the pipit when the latter saved itself by diving deep into a patch of thick heather. The Merlin came back and several times made swooping dives at the spot where the pipit had disappeared in an attempt to flush it. Sometimes this tactic works and the prey is frightened into taking flight again where the raptor has the chance to make a kill, but on this occasion the pipit sat tight. Suddenly the Merlin gave up; half a minute later it was hard on the heels of another small bird away on the far side of the island, a chase which took both birds out of my view.

On the 11th the wind finally moved away from the north-west and veered gradually round to a gentle south-easterly. With some rain overnight there was a substantial fall of night migrants at dawn; 100 Blackbirds, six Ring Ouzels, 50 Wheatears, 30 Robins, 20 Chiffchaffs, the first four Willow Warblers and 60 Goldcrests. The Goldcrests favourite haunt is the single sitka spruce in the garden and three smaller ones a short distance away, and all day long at least a dozen were in the larger tree, their thin 'zit-zit' calls the chief sign of their presence. About mid-morning I heard a slightly different call from the tree, higher pitched and distinctly harsher in tone. I waited patiently, checking every little movement in the branches and after ten minutes the bird I was looking for appeared near the tip of one of the outer branches, a Firecrest. Brighter green above, purer white below with a red crown and a thick white eye stripe and a beautiful copper bronze patch on the 'shoulders' he made the Goldcrests seem drab by comparison. During the day several Sand Martins flew across the island, heading swiftly north towards the mainland. One Swallow, the first of the year, was seen as it hurried through the valley at Gibbdale.

In the afternoon five Black Guillemots were sitting on the sea off Cow Harbour, a spot in which they are often to be found. Occasionally one dived, showing the bright scarlet legs as it disappeared from view. A small number of pairs breed on the mainland nearby, within a mile of the Sound and in contrast to the other auks numbers seem to have been increasing slightly in recent years, not only in the Isle of Man but also other parts of the Irish Sea. They have never yet been known to breed on the Calf of Man itself but there is an excellent chance that sooner or later the Tystie (as it is known in Scotland) will be added to the list of species which have nested here. The nesting spot is normally located in a hole in the cliff or in a tumble of large boulders close to sea level and much of the coast of the Calf fits these requirements perfectly.

By evening the wind had dropped completely and it was sunny and warm. Several Blackbirds were singing around the Glen but the song of the island Blackbird rarely equals that of his mainland counterpart. In the most suitable surburban areas with large gardens Blackbirds can reach a density of one pair to every two or three acres but on the 616 acres of the Calf only between 15 and 20 pairs nest each year. Presumably the reason for the inferior song lies in the fact that there is not the intense competition for territories here which prevails in the more favoured areas. The birds sing less and their song never becomes as well developed as on the mainland.

A thick seamist has spoiled several days and made observations difficult. This lifted overnight on the 19th and with a mild southerly airstream over most of Britain the migrants were quickly moving. Over 100 Willow Warblers, 40 Chiffchaffs and 40 Wheatears were the main dawn arrivals and a Grass-hopper Warbler was reeling loudly from the sedges in the field in front of the observatory at intervals all morning. Several Ring Ouzels were obvious near the head of the Glen, flying wildly away at the least sign of disturbance and four Collared Doves assembled in a neat row along the observatory roof.

During the morning a steady passage of hirundines (a group which largely migrate by day) developed. A continual trickle of Swallows and Sand Martins, with the occasional House Martin, were arriving low over the sea and passing quickly across the island. A few were attracted to the millpond and at one time there was a build-up of birds there, dipping down to drink from the surface of the water. But the urge to push northwards while the good conditions lasted was strong and within a few minutes nearly all had drifted away up the Glen. Later in the day a Black-tailed Godwit was flushed from the rocks near Fold Point, only the second one ever recorded from the island, and two Common Sandpipers were watched feeding among the rock pools close to Cow Harbour. This spot is so named because it was the harbour used whenever cattle were swum across the Sound by the Calf farmers.

With a continuation of the milder weather some of the residents are at last showing signs of nesting. A hen Blackbird has been gathering dead grass for her nest from under the fuchsia bushes and several times disappeared into a thick tangle of bramble nearby. Most of the small passerines usually begin nesting rather late on the island and this is doubtless due to the much greater exposure of the terrain as compared to an inland area. It is June before the cover really begins to thicken and the first nests of species such as Dunnock and Blackbird suffer from a high rate of predation, particularly from the corvids. The Stonechats, however, build their nests deep in heather or gorse and they are invariably early. A few days ago a nest was found among heather in Gibbdale and the hen is now sitting tightly and doubtless has a full clutch of eggs. The Herring Gulls are also beginning to build and along the southern shore where they nest in greatest profusion many are collecting clumps of grass from the sides of the steep grassy cliffs.

The Choughs are more advanced in their breeding cycle and all four pairs are now incubating. On my last visit to Caigher Point I quietly approached the crevice in the cliffs in which the pair there nest and looked inside. When my eyes became accustomed to the darkness I could just make out the form of the bird sitting on the nest four feet away at the back of the crevice. It never moved a muscle and I carefully withdrew. The female incubates the eggs and is largely fed during this period by the male. The latter behaves quite inconspicuously during the incubation, feeding, preening and resting somewhere on the cliff top away from the nest. Now and again the familiar 'kee-aa' calls will ring out across the island and a party of six or eight will join together in the air for a few minutes in one of the social flocks which are so much a part of their life. Choughs do not breed until they are at least two (perhaps three) years old and several non-breeding birds regularly form part of these gatherings.

The other three nests in use this year are all in precipitous cliffs and quite inaccessible. The east coast pair nest in a most fearsome spot. Giau Yiarn is little more than a fissure in the cliffs, thirty feet wide and 300 feet from top to bottom, with sheer sides running with water. Eighty feet down from the top on the eastern face are two small deep cavities and in one of these the Choughs nest, safe from interference from any human predator.

A series of Atlantic depressions from the 22nd brought rain and strong winds and effectively halted migration but a brief halt to this cycle allowed a new rush of migrants and many Willow Warblers were present on the morning of the 28th. Although there were concentrations in the withy and the observatory garden most of them were well spread out in the cover alongside the old field walls. Now and again a short burst of reeling from the thick patches of sedge in the Glen revealed the presence of a Grasshopper Warbler but the birds positively refused to show themselves. A Sedge Warbler singing loudly from the withy was equally elusive but several finely plumaged cock Redstarts were much more easily watched and a male Blackcap spent the day in the single ash tree at the head of the Mill Giau. This Giau is a steep sided gully some 200 yards long which ends in cliffs in the small bay known as the Leodan. The bottom of the gully is well sheltered and the grassy slopes above the stream now hold a mass of soft yellow primroses. The first bluebells are coming into flower among them and it is a spot much favoured by butterflies.

Small Tortoiseshell, Green-veined White and Small Copper are the species most often found there.

The sole pair of Mistle Thrushes which breed have been sitting for several days on four eggs in a nest in one of the sycamore trees but this morning the nest was empty and deserted. A broken eggshell on the ground nearby gave a clue to the reason; undoubtedly the nest had been discovered and robbed by a Magpie or Hooded Crow.

Many of the Herring Gulls have their nests completed although none appear to have yet begun to lay. A few of the Greater Black-backs along the east coast now have eggs although only one nest so far contains the full clutch of three. About twenty pairs nest along this half mile stretch of coast, mostly on the highest grassy hummocks or the edge of the broken, low cliff as they seem to prefer to look down on their lesser neighbours, the Herring Gulls, which nest in much greater numbers on the rocks beneath.

MAY

For the first three days of the month the weather remained cool with a fresh south-west wind. On the 3rd the first two Whitethroats of the year put in a very belated appearance and a Swift was also seen. Before dawn on the 4th there was a very marked change in the weather. The wind had backed to a light southerly and the temperature had risen considerably. At first light the bushes around the front garden and the fuchsia hedge were alive with birds, Sedge Warblers, Whitethroats, Willow Warblers and others, all feeding very actively after their long night's flight. Most night migrants seem to feed during the morning, rest later in the day and the majority will leave at dusk on the next stage of their journey north. In all likelihood these birds had set off soon after dusk the previous evening from somewhere in Wales or south-west England. Most of them had probably been flying for between four and eight hours and had covered something between one and two hundred miles. Ringing occasionally provides us with striking examples of such flights. A Willow Warbler ringed at the bird observatory on Portland Bill in Dorset on 5th June 1963 was recaught on the Calf exactly 48 hours later, a journey of 260 miles. There is no way of knowing of course, whether this distance had been accomplished in one night's flight or two but even a bird as small as a Willow Warbler could certainly make such a flight non-stop provided it had sufficient energy reserves in the form of fat.

On the first drive of the main heligoland trap a dozen birds were caught with ease and for several hours this trap and four mist-nets were sufficient to catch all the birds which we could safely and quickly handle. Each bird caught is placed in a small linen bag and then taken inside the observatory where it is examined, measured, weighed and ringed and then it is released. On most spring mornings the number of birds on the island slowly reduces because some work their way north, cross the Sound and continue feeding on the main island, but there were so many birds on the island this morning that those that left the observatory area were continually being replaced by others that were moving up the Glen. Still more birds arrived along the old field walls which converge on this oasis, from the lighthouse area to the west and from the Heath to the east, so that there was a continual stream of birds through the trapping area.

The ringing went on until dusk and by the end of a busy day 223 birds had been caught. This total included 34 Sedge Warblers, 28 Whitethroats, 84 Willow Warblers, 52 Chiffchaffs, 10 Goldcrests and ones and twos of nine other species. The one Wheatear trapped was a large brightly coloured bird with a long wing and clearly belonged to the Greenland race.

Another large fall of warblers occurred in continuing fine weather on the 6th. Grasshopper Warblers were particularly prominent in this fall and on the short walk to the lighthouse I flushed nine from the bracken and sedge. Several went off low over the ground with a weak, fluttering flight and it was difficult to imagine them completing the journey to Scotland safely, let alone flying all the way from north Africa. Such skulking birds are difficult to count but when we filled in the log that evening we were able to account for at least 75 on the island.

Later I sat on a wall near the withy, which was humming with birds. Several Sedge Warblers sang from the thick cover and were answered by others further afield. Willow Warblers were fly-catching around the withy edge while the Whitethroats 'churred' from bramble bushes nearby. A hen Blackcap showed briefly among the willows; a Redstart flew across the path in pursuit of some insect and a loud bill snapping signalled the appearance of a Spotted Flycatcher. As I walked back up the Glen a Tree Pipit flew overhead, identified by his distinctive buzzing call note. Then a Cuckoo flew off the wall not twenty yards away and I had scarcely moved on when a strange looking straw coloured bird leapt out of the grass at my feet. As it flew off with its long legs dangling it showed a bright chestnut patch on the wing. It was a Corncrake, a bird seldom seen because it is so adept at running off through the grass as one approaches. In the fields at the head of the Glen Wheatears and Whinchats were scattered widely. There were at least 25 of the latter, most of them richly plumaged cock birds. Another 191 birds were ringed that day.

Smaller but still substantial falls of warblers, chats and flycatchers occurred each day up to the 13th. The weather was mostly fine and warm with light easterly winds. By the end of the ten day spell exactly 1,000 birds had been caught and ringed and thousands more had passed through the island in one of the best periods of migration for several springs. Some of the more unusual birds seen included a Lesser Whitethroat, a scarce visitor to the Isle of Man, a Quail flushed one day from the fields near the withy and a male Hen Harrier which spent two days hunting on the island. This was the end of the main period of spring migration although late migrants (such as Spotted Flycatcher and Turtle Dove) and stragglers will be passing for the next month.

On the 14th on the first early morning drive of the main heligoland trap a strange warbler was caught. When I took it from the carrying bag for examination I was completely puzzled as to the birds identity. Superficially it resembled a Chiffchaff and obviously it belonged to the same genus (the Phylloscopus) but the upperparts were a darker uniform brown and there was no trace of yellow or green in the plumage. The underparts were buffish-white and it had a long, broad eye-stripe. The most startling feature was a warm pinkish-buff tinge to the cheeks, throat, eye-stripe, underwing and undertail coverts. This is always a most exciting moment; to see or catch a bird which is obviously unusual and to which one cannot immediately put a name. A description was taken of each part of the birds plumage, details of its wing formula noted and the wings, tail, bill and tarsus measured. After it had been weighed and a few photographs taken in the hand the bird was released. Immediately it dived into the cover of the fuchsia bushes and gave no chance of field views.

The next step was to establish the bird's identity. A first run through the books and guides suggested Dusky Warbler but this seemed extremely improbable because all the eleven or so previously recorded in Britain had been in October or November. Thinking I had made a mistake I checked every species possible with the details which had been recorded. Every one but Dusky Warbler could be confidently eliminated while everything known of the strange warbler fitted this species. It was a unique record. The Dusky Warblers' breeding range stretches from the middle of Asia to the Pacific Ocean while they winter in south-east Asia. A very few have reached western Europe in the late autumn but never before in spring. A few hours later the bird was recaught and on release this time it was more cooperative and allowed itself to be watched for some time as it fed near to the ground. Finally it disappeared once more into the fuchsia.

Some of the most spectacular sea-cliffs on the island are those on the east coast 200 yards either side of Kione Roauyr. Here the cliffs rise sheer out of the sea to well over 200 feet and there are plenty of ledges to provide inaccessible nesting sites for the seabirds. This is the main nesting area for the Fulmars and about thirty pairs nest on these cliffs each year with another twenty pairs along the west coast. The Fulmar in its spread around Britain first bred here in 1936 and the numbers have slowly built up since then to the present level. They are very curious birds and if one pauses on the cliff edge where they are gliding back and forth it is not usually long before several of the birds come in for a really close look at the human intruder. To do this they glide slowly past at cliff-top height, often less than ten feet away and turn their head to look at one 'eye to eye' as they pass, sometimes even hanging on the wind for several seconds at little more than arms length distance. This is also the only spot on the island where Cormorants nest. Although never numerous here they have declined over the past fifteen years and now only about six pairs breed, building their bulky nests on the same ledges which have been used for at least thirty years and probably many more.

Perhaps the most elusive of the birds breeding on the island are the Short-eared Owls. One pair are resident in most years and the nest may be almost anywhere in the large area of heather and bracken on the east side of the island. When feeding a growing nest of young the adults become more adventurous and will hunt by daylight but at least in the early stages of the breeding cycle they can be very inconspicuous. This evening (the 21st) I went up to a high point overlooking the Heath. Eventually when half the light had gone one bird put in an appearance but it was only hunting and gave no clue as to where it might nest. The flight of this owl is beautiful to watch. The down beats of the wings are very deliberate with a distinct pause between each and the flight appears so soft and silent that one has the impression of a huge moth moving over the ground. There are frequent spells of gliding with the wings held in a shallow 'V'. Methodically the bird quarters the ground and shows a surprisingly agility whenever a possible meal is found, twisting and diving into the grass in an instant.

With most birds the adults hunt independently for food when they are feeding young. Not so with the Choughs. Once the young are a few days old both adults go off together to fetch food, another example of the strength of the pair bond in this species. Two pairs are regularly visiting the fields at the head of the Glen and gather much of their food in the short grass on the higher, northern sides of these fields. Both birds, keeping close company, gather insects for anything up to half an hour, swallowing all they find. When both have a sufficient quantity, off they fly to the nest together and the young are fed by regurgitation. As when they are building it is not a very difficult task to trace them back to their nests. The nest on Caigher was recently checked and contained only one young bird. Although they lay a clutch of four or five eggs Choughs often only rear one or two young. Because they nest in such difficult cliffs it is not easy to find the reason for this, but in several nests I have watched, one or two of the young died in the few days after hatching.

While small numbers of Swallows are still passing daily on northward migration the few pairs which breed have been settled on the island for a fortnight and nest building will soon begin. The attraction for them here is the nest sites provided by the old buildings and each year four or five pairs breed, spread between Janes house, the sheds by the observatory and the towers of the old lighthouses. The doors of the old farm sheds are tied open for the summer to give them plenty of choice of nesting sites.

For most other species the breeding season is now in full swing. The few pairs of Lapwing in the Glen and near Caigher have small young and have a difficult task to protect them from the attentions of the Hooded Crows. Several pairs of the latter have young in nests in the cliffs. Two young Ravens have recently fledged and are already strong on the wing. About fifteen pairs of Oystercatchers nest on the lower parts of the coast and their eggs are close to hatching. One or two Shelduck can often be seen in the south of the island. Although they have nested in the past these birds seem mainly to use the island as a safe resting place when off duty. Mallard do breed though and one brood of nine ducklings appeared on the millpond this week. With so many large gulls about their chances of fledging are not at all high.

Large areas of the Glen and the north end of the island are now carpeted with millions of bluebells. This glorious show will not last much longer for the bracken is rapidly pushing up among them and soon it will form a dense mass over these areas. The cliff tops are ablaze with the delicate blue of spring squill and a great carpet of pink thrift is just coming into flower all along the south-west cliffs.

Late May is the time when most Collared Doves are seen on the Calf of Man. Walking back along the path above Gibbdale on the evening of the 26th a flock of eleven flew past me, circled the farm and landed in the sycamores. They appeared to be restless and within a few minutes left the trees and rose high in the air. First they headed south for half a mile, then swung round in a big circle silhouetted clearly against a beautiful mackerel sky and finally headed north until out of sight. The first one was not recorded here until 1962 but now they occur commonly each year between mid-April and mid-June although they seldom appear at any other time of the year. Probably most of the birds which visit the island are pioneers, part of the contingent which are still pushing the range of this species to the north-west.

Very often with birds, memorable incidents occur which occupy the space of no more than a few seconds. I was walking along the path in front of the observatory when my attention was suddenly arrested by a loud rush of wings from a large bird near at hand. Startled I looked up and almost hovering over the bushes not twenty feet away was a female Peregrine. It was looking down intently at something beneath it and for a few moments I was able to take in

every detail of this magnificent bird; the fine barring on the underparts, the
thick black moustache and the bright yellow cere and legs. Suddenly it
spotted me and was gone in an instant, away across the fields. When I looked
round the other side of the bush the object of its attentions, a Turtle Dove, was
lying unharmed in the middle of a mist-net!

JUNE

The only stock now kept on the island are a few sheep of a breed peculiar to the Isle of Man known as Loghtans. This breed is no longer popular with farmers and most of the remaining animals are owned by the Manx National Trust. The Loghtan lambs born this spring are now well grown and as I leant against the wall of their field to watch them playing in the sun I heard a harsh chuckle and caught a glimpse of a large thrush disappearing over the wall of the next field. It flew back onto the wall and sure enough it was a Fieldfare, the first summer one I had ever seen. The last of the birds returning north passed through a month ago, so what was this bird doing on the island when it should have been on its breeding grounds in Scandinavia?

Later, a visitor came rushing into the observatory in a state of great excitement. A few minutes before he had been walking through some long grass when a small bird 'just like a tiny Partridge' had risen under his feet and flown rapidly into the next field. His description was good and undoubtedly he had been lucky enough to flush a Quail, one of the most difficult species to see and one for which most bird-watchers have to wait many years. He had noted the spot where the bird landed and we went back and walked systematically over the ground but of course the bird had run off through the grass and was not to be found.

The main Shag colony is situated on the east coast just past the point of

Kione ny Halbey. Here the ground starts to rise sharply and a path branches off and leads along a grassy slope to the colony. Where the path ends is a broken cliff face at a fairly easy angle with many large and small boulders lying about on the slopes. In the gaps and cavities under and around the boulders are the nests of about eighty pairs of Shags. With care it is possible to walk about on this slope and nearly all the nests can be inspected closely. On the seaward side of the colony there is a ridge of black rocks about twenty feet above the sea and here the non-breeding and off-duty birds line up when they are not out at sea fishing. As the colony is approached the birds on these seaward rocks come into view first. In all, 64 were lined up there and a few had their wings outstretched in the manner of the Cormorant tribe the world over. The birds on the rocks stood their ground until I was close to the first nests when several panicked and within a few seconds they were all airborne and glided down to sit on the sea just offshore. Their departure alerted the birds on the nests to my presence and several shiny dark green heads and necks came poking out from among the rocks to see what had caused the mass departure. On seeing me they set up a loud grunting, accompanied by much head waving and more heads came peering out from under boulders further away to see what was amiss and so gradually the whole colony became aware of the intrusion.

It was easy to see where most of the nests were situated because of the splashes of white droppings on the rocks beneath each nest. I approached the first site and knelt down to look underneath the boulder. At once there was a scuttling sound and as my eyes became used to the dim light I could make out the last of two or three large young disappearing into a crevice some six feet behind the nest. These birds were quite close to fledging and at this age are very mobile and will soon return to the nest. At the next nest the three young were of a similar age but their nest cavity was much smaller and they had no escape route so I was able to take them out one by one for examination and ringing. They were almost as large as the adults but were still covered in a coat of thick grey down. As I reached in for them the loose, flabby skin around their necks was quivering violently and their eyes were wild, giving them a very nervous and frightened expression.

The third nest was in a more open site and contained a sitting bird. Shags are very bold in the defence of their young and I was able to walk up to within three feet of the nest whereupon the sitting bird raised itself onto the rim of the nest, revealing three tiny recently hatched young, and hissed and spat at me with considerable spirit. At a distance an adult Shag appears quite black but at close quarters there is a beautiful irridescent green sheen on the plumage, hence the birds local name of Green Cormorant. A flange of skin around the gape is bright yellow and the eye is the colour of jade. When the young first hatch they are tiny black lumps of naked flesh, reptilian looking and rather repulsive. Invariably there is a marked size difference between young in the same nest at these early stages and this is because the eggs are laid at intervals of a day or more and incubation begins as soon as the first egg is laid. This showed very clearly in one of the next nests, where one egg was chipping, one young bird was just hatched and the other young was almost twice as large already.

Some nests still contained clutches of eggs and from these the bird sitting usually flew off as I approached. These late nests are repeats of earlier ones which have failed and also probably those of younger birds nesting for the first time. The nest is fairly substantial and most in this colony are built largely of dead bracken stems although many also contain seaweed. Frequently they are decorated with a few fresh green leaves from some plant such as sea campion growing on the cliffs. Until the young birds hatch the nest remains quite fresh but as they grow it rapidly deteriorates as their droppings foul the area and becomes a soggy, stinking flattened mass. The whole colony in fact gives off a fairly strong odour and an afternoon spent ringing Shags is a very messy business.

Apart from this large colony there are about a dozen smaller ones around the island, most of them accessible. In all, close to 300 pairs nest annually and they have shown a steady increase in numbers over the past dozen years. In some places on the cliffs there are the scattered nests of Shags preferring to breed alone. Many seabirds, Kittiwakes for example, need the stimulus of a large colony for successful breeding and attempts at isolated nesting are almost always unsuccessful. However, the Calf Shags seem to be equally successful whether they choose to nest solitary or in one of the colonies. It would be fascinating to know what factor determines their choice of nest site. Perhaps, as with humans, some are more gregariously inclined than others?

Some of the summer visitors are surprisingly late in arriving and even now are still passing in small numbers. One or two Spotted Flycatchers arrive daily and this morning (the 11th) there were at least three in the garden and two more in the ash tree in the Mill Giau. Swifts are quite regular in small parties although in their case it is difficult to separate true passage from the lengthy feeding movements they perform in summer. Most days a few House Martins move quickly through the island and one or two Turtle Doves appear. As often as not the latter are flushed from the large sycamore trees in the back garden for they are essentially a tree loving species and seldom pause for long on the open spaces of the island. For three days a Chiffchaff sang regularly from the garden and even at times the chimney of the house but he seems to have moved on now. Perhaps this was a young bird which came north too late to nest and is leading a wandering life until it is time to return south.

Apart from a short gap at each side of Cow Harbour the Herring Gulls nest in numbers around the whole of the coastline of the island. However, the greatest numbers are congregated along the southern shore between South Harbour and Caigher Point. A close approach to this area anytime between April and August sets up hundreds of gulls into the air, all protesting loudly at being disturbed from their nests. Even from the observatory a mile away it is easy to tell when someone is walking in the gull colonies and to see where from the cloud of gulls over their heads. The cliffs in this region are very broken and mostly quite low, and the majority of nests are scattered about among the rocks on fairly level ground. In places, as at South Harbour itself, the nests are situated very close to the sea and occasionally a summer storm coupled with a spring tide washes away some of these. In all about 2,000 pairs currently nest on the island and this is an increase of approximately 500 pairs on the number breeding in 1959. So the rapid increases in the numbers of this gull which are taking place in some parts of Britain are not apparent on the Calf of Man.

Walking among the nests it is noticeable that many are decorated with some man-made piece of flotsam picked up from the sea. Fragments of nylon fishing rope are the most regular items but plastic bottle tops, bits of polythene and even a child's plastic toy soldier are all utilised! Reds, oranges and yellows seem to be the birds favourite colours, unless it is that these bright shades are more easily spotted in the sea than greens, blues and browns. The other gulls and especially Shags also go in for such artificial nest adornment.

Nearly all the gulls nests now contain small young up to two weeks old. When first hatched the chicks are most attractive bundles of soft, grey down. They are active from a few hours old and after a few days usually leave the nest and seek a better hiding place in the rocks or vegetation close by. However it is perilous for them to wander in the colony for if they enter the territory of another Herring Gull they are attacked and sometimes killed by the adults nesting there.

Great Black-backed Gulls nest in small numbers in the same areas as the Herring Gulls. They monopolise the highest points and elevated rocks and often several pairs nest close together as on the top of the Burroo.

In 1791 a Colonel Townley visited the Calf of Man and later he wrote the following in his journal. 'We then went down to the boat and rowed round the greatest part of the island, which presented us with some of the most striking, romantic scenes I ever had met with; and with so great and such a mixed multitude of birds as no one spot in the universe can (I believe) exhibit, for their numbers were so astonishingly great indeed, that I do not know how to liken them, but by scriptural comparison, as the stars in the firmament, or the sands upon the sea-shore, which are beyond the art of numeration.'

The birds which Townley was referring to especially were the auks; Puffins, Razorbills and particularly Guillemots, and while it is evident that there is some degree of exaggeration in his writing it is also abundantly clear that these species have declined drastically since his visit. Now only about 100 pairs of Guillemots and 125 pairs of Razorbills nest on the island and the Guillemots are still declining steadily. The auks have, of course, shown a similar decrease throughout the whole of southern Britain and this decrease is usually attributed to oil pollution. However, I believe that direct human persecution of the birds themselves for the sake of their feathers and flesh and the taking of their eggs for food is a more likely reason for the original decimation of the Calf colonies. There is evidence in old documents dating from the 17th century that such collecting was done annually to a very considerable extent on the island. Such persecution, at least in the form of egg collecting, probably continued into the present century and it is only for the last thirty years that the birds have received effective protection. Given absence from other pressures it is probable that in time the birds could recover their numbers but there is no doubt that oil in the sea is now the grave threat to their continued existence and threatens to finish the job altogether and extinguish these fine birds from the Calf of Man and many other areas.

The Puffins are a particular mystery. In 1907 visitors to the island found them 'numerous wherever the edges of the precipices afforded soil for their burrows'. Yet by the late 1930s few remained and now only between twenty and thirty pairs nest on the steep western cliffs. Exactly when this decrease took place and, more important, the reason why, remains obscure.

One recent morning I was standing on the end of the Burroo Neck watching seals in the clear water below when the pair of Choughs which nest on the huge detached rock, the Burroo, arrived to feed their young. On seeing me they at once began to call loudly and I heard a muffled, answering call from one of the young birds on the rock opposite. Suddenly it flew out from the cave, possibly on its maiden flight for it was extremely shaky and unsure of itself in the air and after a few seconds crash landed onto a steep rock face. The anxious parents flew across to join it and with a good deal of concerned calling managed to persuade it to follow them back into the safety of the cave entrance.

A few days after this incident I was walking towards South Harbour when I heard the loud shrilling cries of young Choughs begging for food. Soon I came on the four birds which were undoubtedly from the Burroo nest, two juveniles easily distinguished by their shorter yellowish-orange bills and orange legs and the two adults which were clearly having a difficult time to satisfy the hunger of their off-spring. I watched from a distance as the adults dug busily in a patch of bare earth; as soon as one paused the waiting young rushed over to it and begged to be fed with loud cries and shivering wings.

Each young was fed in turn from the adults bill. As I approached all four took off together and I was surprised to see how strong on the wing the young birds were already, although as yet they lacked the gracefulness and delicate control which is so characteristic of the adults' flight.

Each year one pair of Moorhens nests on or near the millpond. This is the only apparently suitable habitat for them on the island and as there is plenty of thick cover near the pond they are usually successful and often rear two broods. A second pair have nested in several recent years in a far more unlikely spot, a tiny pool only ten feet wide and six feet across among the rocks above South Harbour. The Herring Gulls nest only feet away and perhaps it is for this reason the Moorhens delay nesting until the sedges around the pool have grown tall and give some shelter. This year's nest contains five eggs, probably not yet a full clutch. The pair appear to live largely on a marine diet for I often see them seeking food among the exposed seaweed at low tide.

A number of other species also show the ability to adapt themselves to rather unusual nest sites on the island. Several times I have found Blackbirds' nests on the ground among dead bracken instead of in a bush and the only Song Thrushes nest I have seen here was in a similar position. In one year a Magpie built a nest in a hole in the cliffs instead of in a bush or tree and the Wood Pigeons which have colonised the island recently usually place their nests on a rocky or earthy ledge, perhaps on the side of the Mill Giau, or even underneath a gorse bush on the steep bank at the side of the observatory. Most unexpected of all was the pair of Spotted Flycatchers which reared three young one year from a nest in the cliff at the sea end of the Mill Giau within fifty yards of high water. Usually this species seeks a sheltered inland garden or wood and they had never previously nested on the island.

JULY

Many visitors are surprised to find Partridges on the island. They are indeed recent colonists, having arrived only in the early 1960s but are now well established and thriving with a population of between twelve and fifteen pairs. At the same time as this increase on the Calf the population over much of Britain has declined seriously. However, perhaps this is not surprising because the general decrease has been attributed to the use of pesticide sprays on farmland, which kill the tiny insects on which the young Partridges feed. Because the island is a nature reserve no chemicals are ever used. The birds also doubtless benefit from the fact that no shooting is allowed and also the absence of ground predators, such as fox, stoat or weasel.

The heather and bracken appears to give them ample cover and provided it is not too wet at the time the young are small they usually rear large broods. There are family parties in several parts of the island at present and the young, although barely half grown, can already fly strongly if alarmed.

The Kittiwakes have an interesting up and down history on the island. Early in this century they used to breed in a colony on the west coast near the lighthouses. This was deserted sometime before 1921 and they didn't nest again until 1956 when thirteen nests were discovered in a gully on the west side of Caigher Point. By 1959 when the observatory was established this colony had grown to 136 pairs. The numbers steadily increased and two more colonies were founded nearby; in Ghaw Lang and in another inlet between this Ghaw and Caigher Point while a few pairs also began to breed on the far side of the island at Kione Roauyr. By 1969, 720 pairs were nesting.

Since then the colonies have taken a marked downward turn. The following year, 1970, similar numbers of birds were present but the breeding season was a disastrous one. Most pairs failed to even build a nest, few eggs were laid and only three young were finally reared. There was no obvious explanation but this kind of failure has been noted in Kittiwake colonies elsewhere and is possibly caused by a shortage of the right food at sea during the breeding season. This year there are about 450 nests but again it looks as if rather few young will eventually fledge. Currently under two hundred of the nests contain eggs or small young. They are late nesters and most of the eggs are laid in mid-June, the young not fledging until mid or late August.

The colony on Caigher Point is ideally situated for observation. The nests are on ledges on the northern wall of the gully and it is possible to sit behind rocks on the south side, only fifty feet away, and watch the birds without causing any disturbance. There are close on a hundred nests and always plenty of activity on the ledges. One of the pair remains at the nest to guard the young against predators such as the larger gulls and, except when out at sea feeding, the other bird of each pair will be on the ledges too. Each time a bird arrives back at the colony from a trip to sea it is greeted by its mate with the well known wailing 'kitti-waak' from which the bird gets its name. The cry is repeated over and over again by both birds and is frequently taken up by other birds round about until half the colony or more is resounding with the rather mournful cry. Then the birds lapse into comparative silence for a few minutes until the next one arrives back when the wailing breaks out once again.

In the middle of this colony is a broad ledge on which a solitary pair of Shags nest each year. They are earlier nesters than the Kittiwakes and the nest now has four fully grown young (a large brood) which will fledge at any time. Why this particular pair of Shags should choose to nest in the middle of a Kittiwake colony and not among their own kind is rather puzzling.

Parties of recently fledged juvenile Wrens have become quite common around the island in the last few days. Holes in the dry stone walls which surround the fields are often used for their nest sites but other nests are built into the sides of steep banks and always well hidden among the surrounding vegetation. For the cliff-nesting Wrens there are a multitude of sites, deep crevices and cavities among the rocks. There is a very predator proof nest in one of the old farm sheds. Above the door, seven feet from the ground, is a small cavity in the brickwork and into this a Wren has jammed its nest of moss. The entrance hole is less than an inch across, and with upwards of five large young packed inside and no capacity for the nest to expand one wonders if the young will fledge before they suffocate.

The location of the Short-eared Owls' nest has finally been discovered. I have been to look for it on a number of evenings over the past few weeks without getting any certain clue as to just where on the Heath it was hidden. On a recent evening I tried a new vantage point and eventually when the light was failing a bird emerged from deep heather to commence its nightly patrol. Returning to this spot with several visitors the next morning we spread out to make a careful search of the ground. Within a few minutes an owl flew out of the heather in front of our line and there at the spot from where it came up was the nest containing one half-grown youngster and a broken egg. It could hardly be called a nest as the bird simply lays its eggs in a depression in the ground among the heather. All around was scattered the remains of their prey; the hind quarters of two rats, part of a young rabbit and the feathers of a Meadow Pipit. As we were examining these grisly remains the adult came silently back across the heather and passed within thirty feet, twice giving a sharp 'bark' of alarm. Before we left the area a quick search revealed another young bird well hidden several yards from the nest. There may have been more; from about two weeks old young Short-eared Owls leave the nest and hide in the surrounding vegetation, a habit which minimises the possibility of the whole brood being taken by a predator. When the adult brings food at night it is able to locate the young because they make a loud hissing noise. As we moved away from the nest area we came on a collection of owl pellets on the ground, probably the spot where one of the pair roosts during the day. I once found the leg of a small bird with a ring still on it inside a Short-eared Owl pellet. The ring had been put on a Meadow Pipit on the island three years previously.

As they grow, Herring Gull chicks gradually lose their early attractiveness until at a month old, just before they begin to fly, they have become awkward looking, gangling juveniles. The nesting colonies deteriorate too, fouled by the remains of young which have died, an accumulation of droppings and the uneaten portions of food brought by the adults. The colonies reek with a rather strong and unpleasant fishy odour but this is typical of seabirds everywhere. Jackdaws seem to find the colonies good places to scavenge and are often to be seen walking about among the gulls snatching morsels of food. Indeed some appear to visit from the mainland solely for this purpose.

The gull chicks are usually ringed when they are quite large, at an age of about three weeks. Out of fright at being handled the chicks frequently disgorge their last meal. Usually this is an evil smelling mess of half digested fish but on one occasion I was ringing a bird when it caused great amusement by depositing on my leg its last meal of bacon rinds and baked beans! Clearly the parent had been foraging on the mainland rather than out at sea.

A few of the young are now beginning to fly and it is always amusing to watch their maiden flights. Usually they take off from the cliff and lumber into the air very shakily, often losing control and dropping a few feet in the air several times. They lose height quickly and come down in the sea not far offshore. At this point they are frequently attacked by a crowd of adult gulls which follow them out from the rocks and dive down on the scared young bird attempting to peck it as they pass, while its parents desperately try to ward off the attacks. All in all it must be a pretty frightening baptism for the youngster. Normally the hubbub dies down fairly quickly and the bird is left alone but occasionally I have seen the attacks pressed home and the bird killed by its own kind.

The period of time between the last of the incoming spring migrants and the first returning autumn birds is very short, little more than a month. Turtle Doves, House Martins and Spotted Flycatchers were passing until mid-June, and by the third week in July the first of the juvenile Willow Warblers are already moving southwards. They are soon followed by several other species. Juvenile Cuckoos are seen most days in the Glen and Common Sandpipers start to appear among the rock pools around the coast. Groups of Swifts pass high overhead, a family party of Whinchats may stay several days near the farm and the occasional Green Sandpiper will feed and rest for a few hours on the muddy fringes of the millpond.

By the end of this month nearly all the Razorbills and Guillemots will have left the cliffs with their young. For the next eight months their life will be spent entirely at sea and the land will not be visited until the next breeding season draws near. The young leave the cliffs when they are between two and three weeks old. At this age they are only about half the size of a fully grown bird. Their wings are not developed at all but they can swim well and have a thick coat of waterproof feathers for protection in the water. Once they reach the sea they are accompanied by one of the adults and it is several weeks before they are fully grown and able to fly.

At the time they fledge young Razorbills especially can be easily confused by the inexperienced with Little Auks, which they resemble in size and plumage. On several occasions I have received reports of Little Auks near the island in July from visitors to the area. The Little Auk is an Arctic species, very unlikely to occur in Britain at this season and undoubtedly these birds were recently fledged young Razorbills. There are even a number of reports of summer Little Auks in the literature which are highly suspect. The field guides generally in use fail to point out the possibility of this confusion.

AUGUST

In a small well defined area like the island any increases in the bird population are especially noticeable, and these are most apparent at the end of the breeding season. In spite of all the predation by various corvids the Meadow Pipits have reared many broods and there are newly independent young ones everywhere now. Twenty or more scattered about on the field walls or on a rocky bank is a common sight. About a hundred pairs breed with a noticeably higher density in the old farmland areas than in the true heathland. The family parties of Linnets are banding together and several flocks are prominent in the south of the island.

The vegetation has now reached its maximum height and the island looks very different when compared to the spring. In many places the bracken grows to four or five feet high and some of these tracts are quite extensive and difficult to penetrate. Wrens and Dunnocks creep about well hidden among the stems and juvenile Stonechats and Reed Buntings pick insects from the topmost fronds. If a bird decides to skulk in one of these thick bracken areas it can be a difficult job to track it down. One early August, a visitor found a Melodious Warbler in a large patch of bracken by the smithy. The bird was most elusive, giving a glimpse in one spot then disappearing down among the stems, and the next glimpse would be twenty minutes later and fifty yards away.

Early in the month the number of Willow Warblers passing through increases daily. On the 6th there was a fall of 80 followed by similar falls for the next four days, the largest one of at least 250 birds on the 9th. This is the peak of autumn movement for Willow Warblers although smaller numbers

will pass through for the next month. Almost all the birds in this early rush
are juveniles only a few weeks old. The adults have to complete their moult
before migrating and will travel slightly later. Although the Swallows may
have just laid their second clutch of eggs in Janes House and the farm sheds,
many of the young from the first broods are already on the first leg of their
long journey southwards.

Another bird for which the breeding season is not yet at an end is the Stonechat. The pair in Gibbdale, whose first nest was found back in April, are now
feeding their third brood of the year. After the usual quiet period during
incubation they have suddenly become conspicuous again and each time I
walk through Gibbdale the male sits up on a prominent perch and 'tacks'
loudly until I leave the area. The quieter hen pauses with a bill-full of green
caterpillars until it is safe to return to the nest. Already they have fledged nine
young from the two earlier nests.

At this time of year the Shags have a habit of assembling in large, closely
packed congregations on the sea. They normally fish in scattered groups or
even singly but on many days during August a dense mass of the birds can be
seen off the south or east coasts and often quite close to the rocks. Usually
the gathering comprises up to 500 birds but I have seen as many as 1,500
packed together in this way.

In the flock all the birds will be facing in the same direction and for a while
they will be content to drift along with the current. Then the birds at the front
of the mass will begin to dive, those behind will swim forward and in their
turn dive until perhaps half of the group is under water. Most of the birds
surface behind the mass and at once swim forward to pack closely again.
Sometimes the whole group will take to the wing and move perhaps a mile
away to a fresh spot. This is not done en masse but starts with a handful of
birds flying off low over the sea. A few more will follow every few seconds
until they are strung out in a great moving line perhaps half a mile long. As
soon as they land the dense pack is again reformed.

It is interesting that such behaviour is only seen for a few weeks at this time
of year. It may be that for this period they feed on shoals of fish which are
best exploited by a large group feeding together.

The Choughs are a never ending source of amusement to watch. Recently
I saw the pair from Caigher Point, accompanied by their one young of the
year, come noisily across the fields in the Glen and perch on the dyke by an
old stone cattle trough. After a minute of continuous calling and wing flicking
one of the old birds hopped down into the trough while its mate kept watch
from the top of the dyke. The bird in the trough fluffed out its belly feathers,
lowered itself into the water and proceeded to have a thorough bath, throwing
a good deal of water over each shoulder with quick flicks of its bill. Emerging

dripping wet it hopped back onto the top of the wall and after another noisy conversation the second adult went down for its bathe. Afterwards, while they both sat on the dyke preening, the young one decided to emulate them and entered the trough. Perhaps the water was a bit cold that morning, or perhaps it hadn't yet learnt the form, for it stood there looking quite lost and obviously undecided as to whether it should carry on. Its dilemma was solved when the pair suddenly flew off and the youngster jumped out of the trough and hurriedly followed them.

The first two weeks of the month were warm and settled but on the 16th an Atlantic low produced the first gale of the autumn. The wind blew hard from the south then swung rapidly round through east to finish up in the north-west. A good many seabirds must have been blown into the funnel of Liverpool Bay because for most of the 17th large numbers of Fulmars, Gannets and Manx Shearwaters were streaming back west past the southern end of the Calf. In a two hour watch in mid-morning 642 shearwaters were counted.

Shearwaters are superb fliers, gliding for most of the time on stiffly held wings in and out of the troughs. As they glide they tilt over from side to side every few seconds so that first the black upperparts are showing then the shining white belly. The birds moving past on this occasion had a strong head-wind to contend with but still made easy progress. When they choose to fly with the wind they can move so fast that it is difficult to follow them should the sea be at all rough or the light distracting, especially as they frequently fly along the bottom of a trough at wave level. Speeds of 60 or 70 miles per hour must be regularly attained by these birds, possibly higher on occasions, and all with effortless ease.

Manx Shearwaters have a long association with the Calf of Man. The species was first described in 1676 by Francis Willoughby from specimens taken on the island and their name derives from this. At that time there was a very large breeding colony here, probably running into many thousands, but unfortunately a large proportion of the eggs and young were taken annually for the sake of their flesh, feathers and oil. Some idea of the toll can be obtained from a passage in Ray's Ornithology. 'They usually sell them for about nine-pence the dozen, a very cheap rate . . .. Notwithstanding they are sold so cheap, yet some years there is thirty pounds made of the young (shearwaters) taken in the Calf of Man.' This indicates a kill of over nine thousand young alone and over many years this toll must have been more than the birds could stand, and the colony slowly dwindled. When rats reached the island after a shipwreck in the 18th century it was probably the last blow to the shearwaters and soon afterwards they ceased to breed.

When the observatory was founded the first warden, Einar Brun, discovered that shearwaters were still visiting the island on suitably dark moonless nights. Although marvellous in the air, shearwaters are rather helpless on land and have to visit their breeding colonies at night, when they are safer from predation by the large gulls. They lay their single egg deep underground in an old rabbit burrow or one excavated by themselves. After a considerable amount of searching it was eventually proved in 1967 that the shearwaters were breeding again, albeit in very small numbers, on the steep grassy cliffs along part of the west coast. It is quite possible they never totally deserted the island but have bred intermittently or in small numbers for the past two hundred years.

Late August is when the island looks at its best for then the heather which covers so much of the ground is in full bloom and the brilliant purple carpet is breathtaking in its magnificence. Ling is the more abundant but it is interspersed with large patches of the even more vivid bell heather. The white variety, considered by many to be lucky, is unusually frequent on the Calf. Bees are abundant at this time, and the fuchsia bushes are ablaze with bright red flowers.

This is an excellent period for migrants also and a good variety of species can be seen on most days. Autumn passage is accomplished at a more leisurely pace than in the spring and some of the migrants are content to remain on the island for several days. The 25th provides a good example of an interesting day at this time of year. In an early drive of a heligoland trap a Barred Warbler is caught for ringing. Small numbers of Tree Pipits and White Wagtails are calling as they fly over the Glen. The bushes around the observatory hold a good sprinkling of Willow Warblers, the first few autumn Chiffchaffs, Sedge Warblers, Whitethroats and a Redstart or two, whilst a careful watch on the sycamores in the back garden reveals several Spotted and two Pied Flycatchers feasting on the abundant insect life there. In mid-morning a Reed Warbler is trapped in a net in the withy. This is in fact the first record for the island and causes more excitement than the Barred Warbler which has seven previous records. A good movement of Swallows is under way with the odd Sand Martin and late Swift also passing. Wheatears abound on the heathland on the west side of the island and a family party of Whinchats is prominent near the millpond. Five Kestrels can be seen hovering over the same field nearby while Ringed Plover, Dunlin, Whimbrel and a Greenshank call in flight as they pass over, and a quick scan of the sea from Caigher Point turns up a passing Arctic Skua.

SEPTEMBER

Just to the west of the observatory the main track splits into two, one fork leading towards the lighthouses and the other through the Glen. I was standing by this fork when I noticed a bird sitting on top of the wall close by and saw to my surprise that it was a Wryneck. Normally these birds are very shy, keep to thick cover and are difficult to observe, but this one was intent on sunning itself and appeared quite oblivious to my presence. I was able to stand thirty feet away and admire its superbly marked plumage which, it must be admitted, gave it excellent camouflage against the grey, lichen covered wall. No artist's plate or photograph can really capture the beauty of this bird's plumage at close quarters. It remained passive for several minutes then suddenly flew off low down the Glen with a strong undulating flight.

Yesterday a visitor who had been walking near Caigher Point reported seeing two badly oiled Guillemots drifting past in the current. It seems likely that many birds have been affected because today (the 7th) at least another fifteen, oiled to some degree, have been noted off the coast. I found six in the vicinity of the Sound. Two drifted by in the main channel of water and were quickly carried by the strong current out of sight to the north-west. Another was in smoother water near the Cletts and from a distance didn't appear badly fouled. However, when I got near I could see its belly and undertail were completely stained black by the thick clogging oil. It was making frantic efforts to clean the mess from its plumage and had succeeded in smearing its head and bill with the dreadful oil.

A fourth Guillemot was ashore by Cow Harbour on the rocks which jut out towards Kitterland. This one was covered from head to foot in a great coating of oil and it looked an absolutely pathetic sight, standing there waiting for death to overtake it. I made my way out over the rocks to try and catch it to put it out of its misery (for birds as badly oiled as this are impossible to save) but long before I got near it slipped into the water and swam off into the current.

338

In all likelihood this is only a minor incident, which will never be reported, and of which the public will hear nothing. Only the big accidents such as the *Torrey Canyon*, or when the holiday beaches are threatened, hit the headlines. But it is probably the accumulation of such small incidents as these which do most damage to seabird populations. These minor oiling incidents are certainly not caused by accidents but are the result of someone's careless, deliberate or negligent action. In my experience they are not at all infrequent in the Irish Sea and several times a season we see oiled birds, or oil is washed ashore onto the rocks. The rocks around the Rarick are still thickly coated with oil which came in last winter, some of it in great lumps as big as footballs. A lot of driftwood, which we gather for our fires, is washed ashore in the bays and it is unusual to pick up a piece which is not oiled to some degree by the time it reaches the land.

The politicians talk and make speeches about cleaning up the environment but while they talk the auks are still dying miserably. It is difficult to believe that this could not be stopped if only Parliament had sufficient will and interest. Any company or individual deliberately discharging oil into the sea should be faced with a massive fine or imprisonment. Clearly there are enforcement difficulties but if the penalties were made severe enough this would go a long way to stopping the pollution. Will there be any Razorbills and Guillemots left by the time some affective action is taken, I wonder?

The concentration of migrants on the island naturally attracts a following of birds of prey and the Calf is an excellent place to see raptors. A ringtail Hen Harrier arrived over a week ago and seems to have taken up temporary residence for it can be watched hunting over the fields in the Glen on most mornings. Kestrels are virtually always present, on a few autumn days as many as eight or ten may be seen and, in spite of the lack of woodland, Sparrowhawks often appear. Peregrines are not unusual and the autumn passage of Merlins is just beginning. On one recent evening five Short-eared Owls were seen at once quartering the Glen; this may have been the family party from the island but more likely one or more migrants were involved.

The passage of hirundines, chats and warblers is falling off now and they are being replaced by Robins, Chiffchaffs, Goldcrests and yet more Meadow Pipits. The Goldcrests have been numerous for the past few days with arrivals each morning of between 30 and 60 birds, and one fall of over 200. The sitka tree in the garden was alive with them on this occasion, up to 40 birds at once exploring the needles for aphids. The outer branches of this tree almost touch the observatory and they can be watched through the bedroom windows from a distance of four or five feet.

Choughs have been present in unusually good numbers lately. On most days the excited gatherings have numbered about 22 birds and on one occasion there were thirty together over the heathery banks behind the observatory. The more Choughs gathered together the more noise and excitement they seem to generate and at present they are the most conspicuous birds on the island. Hardly thirty minutes goes by without the birds joining into a compact flock and performing the most graceful aerobatics, accompanied by a good deal of calling, for minutes on end.

Several times recently I have watched them feeding in a most interesting manner. Normally the Chough digs for insects such as ants or moth larvae, or picks its food off the ground. One day I was watching a party of sixteen feeding among the sparse heather near the highest point. Some were digging in the usual manner but several more were tumbling about over the heather in a most bizarre fashion. When I moved nearer I found that they were actually catching insects in flight in a manner not too dissimilar from that of a fly-catcher. They were surprisingly agile and adept at this for such large birds, following the twists and turns of their prey closely and, when successful, catching it with the tips of the mandibles and then alighting on the ground to eat the catch. Most of the insects they took were probably flies, which were abundant among the heather, but one caught what appeared to be a small bumble bee. Usually the flights were low over the ground and sometimes they half-hopped and half-flew in their pursuit but once a bird chased something more than thirty feet into the air. Subsequently I watched a dozen birds which all fed exclusively in this way for more than an hour.

Although it is well into the second half of September two pairs of Swallows still have young from their second broods in the nest. One nest is in the roof of Janes House and the other in the sheds by the observatory and it will be a few days yet before the young from either nest fledges. This will leave the juveniles remarkably little time in which to learn to fly and feed capably before they must set off on their long journey to southern Africa. The only other birds which still seem to be engaged with nest activities are a pair of Wood Pigeons which have a nest with two young on a ledge on a bank behind the observatory. A sycamore tree gives plenty of shelter to this rather unusual nest site. The squabs are only two weeks old and it will be early October before they are flying.

A continuous spell of moderate east to north-east winds during the last week of the month has produced a batch of interesting records. During the morning of the 24th there was a strong movement which included the first Redwing of the autumn, 50 Song Thrushes, three Ring Ouzels, 300 Meadow Pipits, a late Tree Pipit, six Siskins, 30 Redpolls and two early Bramblings. A Lapland Bunting flew over the observatory calling, and in the early after-noon a flock of fifteen Mealy Redpolls was discovered resting in a large patch of bracken in the Glen. This is the northern race of our own Redpoll and a good example can be distinguished in the field by its paler plumage, whitish rump and two buffish-white wing-bars.

On the 27th I was returning to the observatory from South Harbour where the boatman had landed some stores and, pausing to look into the withy, discovered a Yellow-browed Warbler feeding energetically among the willows. It spent the rest of the day there and everyone staying at the observatory obtained excellent views of this tiny little warbler whose nearest breeding grounds are more than 2,000 miles away in northern Russia. The next morning it had gone but the withy held a Lesser Whitethroat, a scarce visitor to the Calf, and there was a second one near the observatory. The withy bird was noticeably brown on the mantle and was very probably of the Siberian race. At least 15 Siskins and 50 Redpolls were seen, a Jack Snipe flushed from the edge of the millpond, and just before dusk a second Yellow-browed Warbler was found in the bracken at the top of the Glen.

OCTOBER

The continuing east wind brought an interesting bird on the first day of the month. This was a Richards Pipit which was watched for a short while on the heath near the track which runs down to Cow Harbour. This large, long-tailed pipit is a rare autumn wanderer to Britain from its breeding grounds on the steppes of Central Asia. Thus this bird had made a journey of at least 3,500 miles from where it was bred to reach the Calf of Man. Rarities such as this certainly claim a great deal of attention from bird-watchers, an attention which is out of all proportion to their importance in our avifauna. But part of this great interest lies in the fascination of knowing that the bird has made such a tremendous journey and originates from a part of the world little known to the watcher. When I see a rarity such as this I invariably find myself trying to imagine just what the bird's home habitat is like. It would be equally fascinating to know the bird's movements once it leaves the island. Will it manage to regain the Russian steppes in the following spring or will it continue to fly west and perish in the cold waters of the Atlantic Ocean?

October is a month of very heavy migration. The vast majority of the summer visitors have gone, although a few Swallows, Wheatears and Chiffchaffs can still be seen on many days, but their place has been taken by Skylarks, thrushes, pipits, wagtails, Starlings and finches. Whilst most of the summer visitors are night migrants the species passing through, now, travel both by night and day and thus a good deal of the movement can actually be seen taking place. Pipits, wagtails and finches in particular travel largely by day.

A typical spell of heavy October movement began on the 11th as the weather

settled down after a spell of strong south-westerlies and became more favourable for migration. There was an early morning arrival of Song Thrushes,
Redwings, Robins and Goldcrests while the number of Blackbirds present
increased markedly. That evening it was mild and dry with a light north-east
breeze. By 21.00 hours it was clear that a large movement of thrushes was
under way as the 'see-ip' calls of Redwings and the 'tchick' calls of Song
Thrushes could be heard almost continually as the birds passed over in the
darkness. There was no moon and a fairly thick cloud cover so I walked up to
the lighthouse and found, as expected, that a number of birds had been
attracted to the powerful beam of the lantern. When still some distance away
little flashes of light could be seen near the tower as the beam momentarily
lit up a bird. The single beam takes fifteen seconds for each complete revolu-
tion and the birds were circling in the same direction as the light, most of them
well away from the tower. In the dim light conditions it is not easy to tell the
species involved but luckily many of the birds were calling and this gave the
best clue to their identity. Many Redwings were present and lesser numbers of
Skylarks, Song Thrushes, Blackbirds and Starlings. A Wheatear flew close by,
showing its startling white rump, and several tiny shapes hurtling round in the
dark eventually proved to be Goldcrests. They might easily have been taken
for large moths. With the permission of the lightkeepers I climbed the stairs
to the top of the tower. As I stepped out onto the balcony a Redstart which
had been sheltering there flew up from under my feet and a few moments
later I was almost hit by a Ring Ouzel which came right in and fluttered against
the glass of the lantern. Often birds are so mesmerised by the light that they
take no notice of humans at all. Many moths had also been attracted and were
clustered on the glass. Angle Shades and Silver Y's in large numbers and a few
Brindled Ochres and Lunar Underwings. Several times I had a glimpse of a
large bird and eventually realised that it was an owl, probably a Short-eared,
hunting small birds in the light's rays. Many lighthouses attract birds in this
manner and a few habitually cause large numbers of casualties. Fortunately
the Calf lighthouse does not come into the latter category and it is very seldom
that a bird is killed here.

At first light the next morning the island was alive with birds. There must
have been thousands of Redwings in the vicinity. A large excited flock came
down into the sycamore trees for a brief rest and there were groups spread out
to feed over the fields in the Glen. For several hours flocks of varying size were
passing overhead, flying west towards Ireland. Song Thrushes and Blackbirds were also numerous on the island. Later I heard from a friend on the
mainland that at least a thousand Redwings had spent the day in fields near
his house three miles from the Calf and he had seen many other flocks in the
southern part of the Isle of Man.

There was a further similar attraction at the lighthouse the next night and many birds were on the island the following day. This time the Redwings were much reduced in numbers but there were twice as many Blackbirds as before. The morning was clear and sunny and the Mountains of Mourne were clearly visible on the Irish coast giving perfect conditions for visible migration. There was a continual stream of birds overhead which built up to a peak by about 09.00 hours and then gradually slackened. Skylarks and Chaffinches were the most numerous with a good sprinkling of Bramblings mixed in with the latter. Parties of Greenfinches, Goldfinches, Linnets and Redpolls passed in lesser numbers and the stream was swelled still further by Golden Plovers, Pied and Grey Wagtails, Siskins, Reed Buntings, and once a flock of 20 Twite. Most of these birds were below 200 feet but some flocks were much higher and in all likelihood there were more, too high to be seen, for birds are known regularly to fly at great heights when on migration. Some of the flocks paused awhile in the trees and bushes around the observatory and parties of Linnets, Goldfinches and Redpolls were feeding on the abundant seeds of thistles, ragwort and burdock nearby.

I was counting the number of Wrens which seemed to be unusually high when I flushed a Long-eared Owl from an elder bush at close range. It flew not forty yards and perched on a rock in the open, giving a splendid view with its ear tufts fully erect. A few distant calls were all that marked the presence of a large excited flock of more than 150 Jackdaws high overhead. They moved westward but within fifteen minutes streamed back in a long line to the mainland. They seem to suffer from a restless urge to migrate at this time of year when the weather is fine but the urge apparently only takes them as far as a good look at the sea and then they return. Titmice seem similarly affected and five Blue Tits were the next arrivals. They spent an hour noisily in the few suitable trees and then vanished, almost certainly back to the mainland.

Next a loud 'cheeping' from the observatory roof arrested my attention. It took a while to place the sound which came from none other than three House Sparrows, the first recorded on the island since April! This is a species very dependent on man, of course, and when the island was farmed they bred commonly, no doubt feeding largely on the grain put down for the farmer's chickens. They left the island when the last farmer moved out in 1958 and since then have only been recorded occasionally in spring and autumn, apart from one year when a single pair stayed to breed. To complete an interesting day a Snow Bunting flew across the Glen calling loudly, at least three Merlins were watched hunting and a Black Redstart was discovered in the lighthouse garden.

Soon after dawn on the 24th it was evident that a large movement of Fieldfares was taking place. A loose party of about 300 flew across the field in front of the observatory heading west, followed by several smaller groups. I quickly followed them along the track to the lighthouse to watch their passage out to sea. The wind was west, force 4 on the Beaufort scale, and the visibility excellent. Every few minutes I would hear the familiar 'chack-chack' and a flock of anything between 20 and 500 Fieldfares would appear, flying low across the island. Mixed in with the flocks were smaller numbers of Redwings while parties of Skylarks, Starlings, Chaffinches and a few Bramblings were passing independently.

On reaching the coast the birds rose steadily higher and higher and without hesitation flew straight out west towards Ireland. At this point the Irish coast is 35 miles distant, perhaps two hours flying time with the fresh headwind which was blowing. Many of the flocks spread out along a broad front as they went out to sea and one group of 300 was scattered across at least a quarter of a mile. From most of the flocks a handful of individuals, not having sufficient strength to attempt the crossing, would detach themselves and drop down into the heather and grass near the cliffs. Some of the larger flocks could be followed out to sea for two miles or more and were still gaining height when they were lost to view, perhaps 2,000 feet above the sea. Soon after 10.00 hours the movement ceased. Over 5,000 Fieldfares had passed through the island in two hours. About a hundred were left on the island for the rest of the day, feeding greedily on the plentiful blackberries.

The 27th dawned grey and dull with a strong south-east wind. Small parties of Fieldfares and larger groups of Starlings and Chaffinches were pushing west in the face of the poor conditions. At 11.00 hours it began to rain and the visibility closed right down to less than a mile and in the next hour many migrants were grounded on the island. Little groups of Chaffinches kept arriving and sought shelter in the trees and bushes around the observatory. Soon there were 400 or more, all 'pinking' vigorously at first from the cover, then many dropped down into the garden seeking food. Lesser numbers of Greenfinches, Goldfinches and Bramblings accompanied them. Parties of Fieldfares and Redwings came low across the Heath and dropped into the dripping wet fuchsias until this little area was seething with birds. Many of the thrushes were bathing in the puddles of water formed on the track.

At 15.00 hours the rain stopped and there was a sudden clearance as the cold front passed over. The wind swung quickly round to a light westerly and the visibility improved considerably. The affect on the birds was quite dramatic. The Chaffinches stopped feeding, flew into the tallest trees and after ten minutes of solid 'pinking' took off in a large flock and headed north for the mainland. Most of the thrushes also quickly departed. The clearance was the signal for the start of a large Starling movement. Within half an hour the first large group of several hundred came rushing through and in the three hours until dusk at least 6,000 passed rapidly across the island and out over the sea towards Ireland.

This movement continued for several hours from first light the next morning. Every few minutes a flock would appear from the direction of the Sound, the smallest only a handful of birds, the largest one of at least 1,500. Each flock followed exactly in the path of the previous one, flying across the field in front of the observatory and passing out to sea just south of the lighthouses. There was an almost military precision about their passage, each flock low and compact as it crossed the island and rising high and spreading out on a broader front as they moved out to sea. Flocks of Fieldfares and Chaffinches were also passing through (the latter in large numbers), but the Starlings were travelling at least twice as fast and I was left with the clear impression that they knew precisely what they were about and where they were heading. The larger the flock, the more compact it was, and the faster they appeared to fly. How they avoid wholesale mid-air collision with one another is very hard to understand.

Suddenly I saw one large flock rise up into the air as it crossed the Heath and wheel about in a most crazy fashion. The reason was not hard to find, a large female Peregrine gliding along high above them. The whole flock was terror stricken and turned this way and that seeking escape, wheeling about in the sky, first in a solid bunch, then stretched into a long line with a few stragglers desperately trying to get back with the flock. The Peregrine followed their every turn and at one point went into a stoop, diving vertically past the tail of the flock and then in the next instant rising high above them again on the momentum gained from its plunge. Clearly to the latter it was just a game for it could have picked off one of the outer birds with the greatest ease. Eventually it lost interest and turned away south and the flock quickly resumed its westward flight.

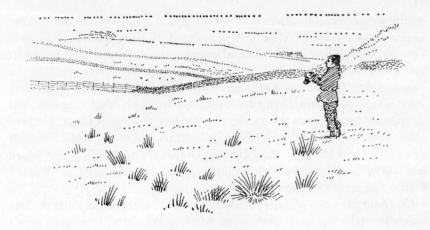

NOVEMBER

A light westerly breeze on the evening of the 8th gave no hint of the storm which was to come. By the middle of the night the wind had backed south and increased rapidly to a roaring gale. At the peak of the storm in late morning on the 9th a tremendous sea was breaking on the exposed southern shore and sheets of wind driven salt spray made it impossible to approach the coast closely. For much of the day it rained from thick, grey, lowering clouds.

Scarcely any birds were to be found, it was as if they had all been swept clear of the island by the severity of the wind. The few seen were down at ground level, not daring to fly for more than a few feet low over the grass. A handful of Blackbirds were turning over the dead leaves underneath the sycamores and a Wren scuttled along the turf into cover. A Song Thrush protested loudly at being disturbed from his shelter under the fuchsia hedge, and a Starling which was foolish enough to take flight was swept wildly up into the air and had to fight its way back to a safe perch.

The north end of the island had the best shelter and several hundred gulls were congregated on the rocks there. Twenty-two Curlew probed the grassy slopes above Cow Harbour and a Whimbrel flew past, his whistle just audible above the wind. Even the seals seemed put out by the storm and six were right inside the harbour, within a few feet of the landing stage to obtain the maximum shelter.

The storm continued throughout the following day, the wind gradually veering round to the west and the morning of the 11th was clear and sunny with a very strong westerly still blowing. A tremendous number of seabirds had been displaced by the gale and all day long there was a heavy southerly passage

347

off the west coast. The strength of the wind was forcing the birds onto the west coast of the Isle of Man and many were very close in to the shore near Cow Harbour. There was a continual stream of Kittwakes passing, some half a mile out but others almost over the rocks. Many were young birds of the year showing thick black forewings. Parties of Razorbills and Guillemots were in view continually, flying low over the waves with rapid wing beats and disappearing for seconds at a time among the deep troughs. A Great Northern Diver flew past followed by a stream of Fulmars and then a Great Skua not a hundred yards offshore. Many Herring Gulls were flying south, also, which was distinctly unusual and their numbers included one immature Iceland Gull, a first record for the island.

One hour in mid-morning proved most exciting with several skuas of three different species (Great, Pomarine and Arctic), a flock of 15 Teal, four Leach's Petrels and a Grey Phalarope. Fortunately, the latter two species were very close in, within a hundred yards of the rocks, for it would have been impossible to have picked out such small birds at any distance over the turbulent sea. Both are oceanic species seldom seen from the coast unless driven in by a severe gale. Perhaps more were passing too far out to be seen. The Kittiwakes and auks continued to fly south in a never ending stream and many thousands had passed before dusk came.

The Choughs are less in evidence now than at any other period of the year. Although they can still be seen daily on the island much of their time is spent on the mainland where they seem to prefer to feed at this season. In particular, much of their day is spent along the shore, where they turn over the rotting piles of seaweed in search of the abundant insect life to be found. The mile long sweep of Langness Bay, six miles east of the Calf, is a favourite winter haunt and up to a dozen birds can usually be found there, feeding close by Redshanks, Ringed Plovers, Dunlins and other shore birds.

Without fail any Choughs which have been on the mainland come back to the island before dusk to roost. One recent evening I watched them returning across the Sound. The first pair appeared with an hour of daylight still remaining, covering the four hundred yards of water with an easy, bounding flight which almost appeared lazy. Twenty minutes later two pairs came over in close company and it was half dark by the time the final five birds made their noisy appearance. All eleven flew down the west coast to roosting spots in the cliffs.

There is still a good deal of migratory activity throughout November, though on a lesser scale than in the previous month and with mainly the same species involved. Nearly all the Fieldfares and Redwings have passed through by mid-month but Blackbirds are still moving and fresh arrivals of fifty or more birds occur on many days. Occasionally a Blackcap or Chiffchaff appears in the now leafless fuchsia (the severe gale of a week ago stripped off every remaining leaf) and probably these are birds which will winter in southern Britain or Ireland. On fine days the finches are still moving through in some numbers and Goldfinches are often quite numerous. A colder spell will bring a few large flocks of Wood Pigeons moving west, Snipe and Woodcock to the moister spots and perhaps a few Snow Buntings flying over.

When the young Fulmars fledge in August the adults leave the ledges and disappear out to sea. During September they are common offshore but these are doubtless birds from colonies further north. Throughout October they are absent from the area and very seldom sighted but about the second week in November they suddenly become common again at sea. Shortly after this the birds return in force to the breeding cliffs, although it will be fully another six months before any eggs are laid on the ledges where they breed. From now on attendance at the cliffs is continuous, the birds spending hours of every day either sitting on the ledges or gliding backwards and forwards along the cliffs.

DECEMBER

Each autumn there is a small passage of Stonechats through the island. Some of these birds travel as far as west France and northern Spain, where two Calf-ringed birds have been recovered, but part of the population is sedentary and several pairs remain on the island for the winter. These pairs are consistently found in the same small areas and this is very likely where they will breed next spring.

Given a normal, mild winter the sedentary birds will probably survive without difficulty. But the hard winter of 1962–63, which was the worst in Britain for over two hundred years, completely destroyed the breeding population on the island. Before that winter twelve to fifteen pairs bred annually; in the two summers following not a single pair nested. One pair returned in 1965 but the population was extremely slow in building up again and by 1970 only four pairs were nesting. This was in marked contrast to most other species affected by that winter (for example Wren and Blackbird) which had fully recovered their numbers within three or four years. However the last two summers have seen a very welcome increase and ten pairs nested this year so, provided no more hard winters occur, it seems likely they will be back to full strength within one or two more summers. The part of the population which migrates south is more likely to survive a severe winter and this may be an important factor in the survival of the species.

The resident pair of Ravens is always conspicuous on the island and especially so at this time of year when Ravens indulge in a good deal of display flying. They are, of course, among the earliest of nesters and provided the weather stays mild will be repairing their nest on the cliffs at Baie 'n Ooig by

February. Both birds soar and glide over the cliffs along the west coast for long periods and the remarkable evolution in which they turn over onto their backs in flight is frequently performed. This trick is done with astonishing agility for so large and heavy a creature. The bird is gliding along in level flight when with a quick sideways flip it turns completely upside down. In the next instant another quick flip brings it right way up again. Usually the manoeuvre is performed several times in fairly quick succession.

One of the island birds, I think the male, has a peculiar call which I have not heard from other Ravens. This is a clear, far carrying 'poo-poo-poo'. When I first came to the island I heard this cry in the distance one day and was completely puzzled by it, and even suspected that perhaps a Hoopoe was on the island, for it is not at all unlike the call of that bird. Usually this call is given when the bird is performing its display flight.

In the bay at South Harbour a Robin is singing from his perch on a jumble of yellow, lichen covered boulders within a few yards of a crashing sea. His song is only just audible above the sound of a heavy swell breaking on the rocks. His companions in this bay are Rock Pipits, Wrens and Dunnocks. It seems a strange spot for a Robin compared to the sheltered surburban garden with which he is normally associated, but this is where this bird has chosen to spend the winter, a long way removed from man and his influence.

And so the year has turned full circle and only the winter birds remain on the island. The migrants have come and gone in their many thousands. Several thousand pairs of birds have bred on the island and they and their progeny have mostly dispersed to places which, in some instances, are thousands of miles apart. Now the Purple Sandpipers have returned to the rocky shore, the flocks of Twite to the exposed cliff tops and a few Woodcock to their favourite hollows. There will be little movement for the next three months, perhaps the occasional arrival of a few winter thrushes or a sudden movement stimulated by the onset of a cold snap. Curlews, Choughs and other birds will come and go across the Sound and perhaps a wintering Hen Harrier will visit to quarter the Glen.